Other Effective Area-based Conservation Measures (OECM) in Marine Capture Fisheries

_____

Other Effective Area-based Conservation Measures (OECM) in Marine Capture Fisheries: Identification, Use and Performance Assessment

---

Serge M. Garcia

Fisheries Expert Group of the IUCN Commission on Ecosystem Management (IUCN-CEM-FEG)

Jake Rice

Fisheries Expert Group of the IUCN Commission on Ecosystem Management (IUCN-CEM-FEG)

Scientist Emeritus, Department of Fisheries and Oceans, Ottawa, Canada and IUCM-CEM-FEG

Anthony Charles

Fisheries Expert Group of the IUCN Commission on Ecosystem Management (IUCN-CEM-FEG)

School of the Environment and Sobey School of Business. St Mary's University, Halifax, Canada

Daniela Diz

Fisheries Expert Group of the IUCN Commission on Ecosystem Management (IUCN-CEM-FEG)

The Lyell Centre, Heriot-Watt University, Edinburgh, UK

CABI is a trading name of CAB International

CABI
Nosworthy Way
Wallingford
Oxfordshire OX10 8DE
UK

CABI
200 Portland Street
Boston
MA 02114
USA

Tel: +44 (0)1491 832111
E-mail: info@cabi.org
Website: www.cabi.org

T: +1 (617)682-9015
E-mail: cabi-nao@cabi.org

A catalogue record for this book is available from the British Library, London, UK.

ISBN-13: 9781836990864 (hardback)
          9781836990871 (ePDF)
          9781836990888 (ePub)

DOI: 10.1079/9781836990888.0000

Commissioning Editor: Jamie Lee
Editorial Assistant: Theresa Regueira
Production Editor: James Bishop

Typeset by Exeter Premedia Services Pvt Ltd, Chennai, India
Printed in the USA

Contents

————————

Foreword

As the Executive Secretary of the first regional fisheries management organization to adopt an other effective area-based conservation measure (OECM), I am very happy to be able to write a foreword to this new book, written by the foremost experts in this new field.

Fisheries management authorities have long underlined that their activities, while focused on harvesting food and economic resources from the sea, can also take into account conservation of biodiversity and ecosystems on which these resources rely. As this book very helpfully sets out in detail, the Convention on Biological Diversity made a pioneering decision in 2010 to acknowledge that sectors other than conservation can indeed provide such benefits. This was not an uncontroversial position for some, given accusations that somehow this was an opportunity for 'greenwashing'. As this book clearly demonstrates, OECMs offer quite the opposite to greenwashing – they can provide rigorous measures, usually already implemented and enforced, that can demonstrate long-term reductions in pressure and impacts on biodiversity and ecosystems, and therefore long-term outcomes for conservation.

In this book, the reader will find a wealth of information on the development of the OECM concept and its accompanying guidance. This is important context to help managers and scientists understand the subsequent detail in the book to guide them in progressing to identify, implement and monitor practical OECMs. All the steps necessary to progress with the identification of an OECM and its integration with other sectors and conservation measures are set out in the book. Indeed, we have used the frameworks as described in this book in the development of the NEAFC OECM. Furthermore, the need to broaden and integrate fisheries OECMs across sectors and seascapes is well described. This is something that NEAFC has been eager to demonstrate in its work, collaborating closely with the Regional Seas Convention in the North-East Atlantic (OSPAR) in the development of MPAs and OECMs. OSPAR and NEAFC have described the 3D interaction of their measures in a joint narrative available on their websites.

Finally, I am very happy to see that the understanding of the principles and aims of OECMs and the guidance are also set out by the authors in a way that should help managers to approach the concepts flexibly to deliver the spirit aimed at in the 2008 CBD decision. This is to enable fisheries management to deliver conservation with a balance of positive environmental, economic and social outcomes.

Darius Campbell
Executive Secretary, North East Atlantic Fisheries Commission

Preface

OECMs offer a tangible area of collaboration between biodiversity conservation and fisheries management, with shared goals, and actions are already taking place. (Joseph Appiott, CBD Secretariat; Opening of the WKTOPs meeting, 2021)

Abstract

This book has been prepared with a particular focus on OECMs established in the marine capture fishery sector (fishery-OECMs). The book has three main goals: (i) to assist fishing nations and authorities in mainstreaming fishery-OECMs, as foreseen in Decision 14/8; (ii) to inform all actors of the fishery sector and all those concerned with marine conservation about the implications of OECMs for marine capture fisheries; and (iii) to help formulate or contribute to the formulation of sound sectoral guidance at regional, national and local levels. The practical implementation of OECMs in marine capture fisheries requires 'translation' of the generic CBD guidance into specific, fisheries-oriented guidelines, reflecting the sector's particular technologies, impacts on biodiversity, governance systems, legal framework, relevant jurisdictions[1] and management regulations. To this end, the book provides detailed technical and scientific background for the identification and use of OECMs in the fishery sector at regional, national and local levels.

We authors of this book have been involved in, and captivated by, the emergence of other effective area-based conservation measures (OECMs) and their role in marine areas of the world. OECMs arose as an outcome of extended negotiations during the 2010 Conference of the Parties to the Convention on Biological Diversity (CBD COP). The area-focused Target 11 in the Strategic Plan for Biological Diversity 2011–2020 included formally protected areas from the outset but there were also demands that the many other types of area-based measures established to conserve and protect biodiversity should be recognized in the target. Nearly 2 weeks of discussions were held to create an approach sufficiently inclusive to satisfy buy-in, yet stringent and explicit enough to satisfy concerns on effectiveness. The expression 'other effective area-based conservation measures' finally obtained consensus and Target 11 was finalized, but the 'other area-based measures' concerned were not sufficiently defined. The efforts of the CBD Secretariat, Subsidiary Body on Scientific, Technical and Technological Advice (SBSTTA) and Conference of the Parties continued between 2010 and 2018 to determine a suitable definition, principles and criteria for OECMs as well as guidance for their governance. Consensus was reached in 2018 in CBD COP Decision 14/8. Since then, work has continued to solidify the idea of and approach to OECMs in the terrestrial and marine domains.

The book was developed between 2016 and 2024 by members of the Fisheries Expert Group (FEG) of the IUCN Commission on Ecosystem Management (FEG), in collaboration with staff members of ICES, FAO and the CBD Secretariat, with the co-ordination of EBCD. In 2016, FEG collaborated with the CBD Secretariat and FAO staff on improving progress reporting and working towards implementation of Aichi Biodiversity Target 6. In 2017, the same team of collaborators worked on the issue of mainstreaming biodiversity concerns in fisheries. By mid-2017, the collaboration moved to experiences on achieving Aichi Target 11 in marine and coastal areas, including the contribution of other area-based conservation measures (OECMs). In September 2017, FEG was asked by the CBD Secretariat to write a report on the use of area-based fisheries management measures, in the perspective of ongoing discussions on Aichi Target 11, and more specifically on OECMs and fisheries (Rice *et al.*, 2018). That report, together with participation in two 2018 CBD expert workshops on OECMs, triggered a 6-year involvement of FEG in fishery-OECMs in a close and fruitful collaboration with FAO, CBD and many other colleagues and institutions.

This book itself grew from successive documents on area-based fisheries management measures and OECMs in marine capture fisheries (Rice *et al.*, 2018; Garcia *et al.*, 2019, 2020a, b, 2021, 2022). The material evolved with time, benefiting from the debates conducted in many different meetings and from scientific articles written by the authors of this book and their collaborators (Garcia *et al.*, 2022, 2023; Himes Cornell *et al.*, 2022; Rice *et al.*, 2022; Petza *et al.*, 2023).

The book draws strongly on the relationship between fishery-OECMs and area-based fisheries management measures (ABFMs), a large range of which are used in fisheries management (Rice *et al.*, 2018; Garcia *et al.*, 2024). ABFMs have aimed mainly at optimizing fish harvesting and maintaining resources and their habitats, but progressively, since the adoption of the precautionary approach to fisheries (EAF) in 2001 at FAO, and particularly since the United Nations General Assembly Resolution 61/105 on Vulnerable Marine Ecosystems (VMEs) (United Nations, 2007a), many ABFMs aim also to protect habitats and a broader range of biodiversity features.

This book focuses on special ABFMs which have been granted the additional label or status of 'other effective area-based conservation measures' (OECMs) when meeting the appropriate CBD criteria, These 'fishery-OECMs' are established by fisheries authorities in fishing areas where fisheries' impacts on biodiversity are significant if not dominant, and the special regulations applied in fishery-OECMs can have a positive and lasting impact on biodiversity. Areas in which fisheries are limited or prohibited by other sectoral authorities (e.g. around oil rigs or renewable energy platforms), or areas already designated as MPAs (including multiple-use MPAs), are not considered in this book, although some of the former may meet the OECM criteria, depending on how the appropriate management authorities have designed and managed their exclusion areas.

This book is not normative but explanatory and advisory. For example, in order to illustrate approaches and lessons learned, it describes specific actions taken or under consideration by many States, RFMOs and local authorities, but these actions are not presented as the sole or best possible way forward on OECMs. The book reviews and provides an interpretation, for the marine capture fishery sector, of the information contained in international instruments which are, in principle, applicable to all relevant socioeconomic sectors. It provides additional considerations from a marine capture fisheries angle, as well as examples of ongoing implementation, that can help to clarify possibly unclear aspects in the CBD Decision or aspects that leave room for adaptation or interpretation.[2] The final interpretation of the CBD Decision, however, remains a responsibility of the State or any other legitimate authority established or recognized by the State for OECM implementation. Indeed, the book's intent is in line with the CBD Decision 14/8 and its guidance on OECMs, the only legal foundation of this instrument. Consequently, it refers to the CBD Strategic Plan for Biodiversity and its Aichi Target 11 on conservation measures, including OECMs (CBD, 2010). It also refers systematically to CBD COP Decision 14/8, the OECM definition it adopted, the criteria and principles it proposed for their identification, and the voluntary guidance it contains on spatial conservation measures (CBD, 2018). Finally, the book recalls the CBD Kunming-Montreal Global Biodiversity Framework (GBF) and its Target 3 – the successor to Aichi Target 11 – on conservation areas, including OECMs (CBD, 2023c).

The literature available on fishery-OECM implementation, in general as well as in fisheries, is still evolving and fisheries are still learning by doing. Many valuable lessons have already been taken from experience, but the suggestions in this book need to be further tested on the ground, in different contexts, to be progressively improved once a meaningful amount of experience and elements of 'best practices' for the fishery sector have accumulated. Progress on fishery-OECMs is being made in regional fisheries management organizations (RFMOs) like NAFO and NEAFC[3] and it can be hoped that, as often expected, regional organizations will help trigger faster and more coherent actions at national level, contributing to strengthening global conservation networks in national jurisdictions and beyond.

There is a broad and diverse target audience for this book. Fishery-OECMs are management instruments bridging sustainable uses of biodiversity by fisheries with conservation. The evidence providing the basis for selection, planning and management of fishery-OECMs includes the best science available and the knowledge of fish harvesters, processors and managers, Indigenous Peoples and local communities.[4] The book should be of interest and accessible to all these actors, including policy makers, managers, scientific advisors, fisheries and conservation scientists and sector representatives. It may be used as background in local, national, regional or larger-scale meetings, aiming at informing about fishery-OECMs and promoting their assessment, use and performance assessment. The document therefore caters to a large readership in regions ranging from well resourced to extremely limited in data and quantitative competences.

The structure of the book follows the sequence of phases required for the full OECM implementation cycle. After a brief introduction on the history of OECMs, the book describes CBD Decision 14/8; the OECM implementation cycle; the enabling frameworks for such implementation; the OECM identification process; their integration in fishery systems and in the broader seascape; their monitoring, performance assessment and reporting, at national and international level; and their eventual revision process. Specifically, Chapter 1, the introduction, provides a short history of the emergence of OECMs in conservation and fisheries management and the events leading to this book. Chapter 2 reviews the foundations of OECMs provided by CBD Decision 14/8, their definition, identification criteria and general principles. Chapter 3 briefly describes the full implementation cycle of OECMs from initial knowledge-based identification to final reporting at international level. The following chapters follow sequential phases of that implementation cycle. Chapter 4 stresses the need to align the national and fisheries enabling frameworks necessary to ensure a successful implementation process. Chapter 5 reviews the fundamental identification process based on the set of adopted criteria, from an initial quick screening through a full assessment by assessors, to their formal recognition by the legitimate authorities. Chapter 6 considers the practical integration on the newly identified fishery-OECMs management into the management plan of the fisheries in which they operate. Chapter 7 describes the monitoring process necessary to undertake the recurrent performance evaluation required by Decision 14/8 to maintain the OECM status. Chapter 8 looks at the potential revision of an OECM when its status indicates an insufficient performance.

Notes

[1] A 'jurisdiction' is the power, right or authority to interpret and apply the law. The term is also often used to refer to the areas in which these authorities operate. The authority having jurisdiction in fisheries and/or biodiversity management may be international, national or subnational; public or private; exclusive or shared.

[2] As formally allowed by the Decision XIV/8 call for a case-by-case flexible implementation.

[3] NEAFC and NAFO have registered respectively 20 and six OECMs, mainly protecting vulnerable marine ecosystems.

[4] The Convention on Biological Diversity does not define the expression 'Indigenous Peoples and Local Communities'. The United Nations Declaration on the Rights of Indigenous Peoples (UNDRIP) (United Nations, 2007) does not adopt a universal definition for 'indigenous peoples', and a definition is not recommended (see the 2018 CBD COP Decision 14/13). Nonetheless, without controversies, Indigenous Peoples and other local communities have been given significant recognition in CBD Decision 14/8 and later in the GBF, recognizing their role in biodiversity conservation.

Acknowledgements

The elaboration of an early draft of this book was supported by the Nordic Council of Ministers, Fisheries and Oceans Canada, the European Board of Conservation and Development (EBCD) and the Fisheries Expert Group of the IUCN Commission on Ecosystem Management (IUCN-CEM-FEG).

The authors are grateful for the numerous and valuable comments received from participants during the scientific workshops organized around OECMs in marine fisheries between 2018 and 2021, for which earlier versions of this document were used as background and basis for discussion. In particular, our most grateful thanks go to the ICES staff and participants that have contributed so much to the success of the 2021 WKTOPS ICES-FEG workshop and particularly Mark Dickey-Collas (Chair of the ICES Advisory Committee) and the co-chair of the workshop (Ellen Kenchington, Fisheries and Oceans Canada) and the leaders of the break-out groups who, through their detailed comments, provided the incentives to further develop the background document of this workshop into this book: Andrew Kenny (CEFAS Lowestoft Laboratory, UK); Marty King (Fisheries and Oceans Canada, Maritimes Region); Llucia Mascorda Cabre (University of Plymouth, School of Marine Science and Engineering, UK); David Miller (ICES); Francis Neat (World Maritime University); Emma Sheehan (University of Plymouth, School of Biological and Marine Sciences, UK); and Peter Wright (Marine Scotland Science, UK).

The WKTOPS background document benefited also from valuable contributions, comments and advice received from the IUCN-CEM-FEG and more particularly: Johann Augustyn (South Africa); Serge Beslier (France); Hugh Govan (Fiji, UK); Alf Hakon Hoel (Norway); Michel Kaiser (UK); Eskild Kirkegaard (Denmark); and Merle Sowman (South Africa).

We benefited also from advice and comments from Joseph Appiott (CBD Secretariat); Gunnstein Bakke (Norway); Kim Friedman (FAO Fisheries and Aquaculture); and Amber Himes-Cornell (FAO Fisheries and Aquaculture); and Dimitra Petza (Greece).

Finally, we are honoured to have direct contributions to the book in the form of boxes authored by Brynhildur Benediktsdóttir (NAFO Executive Secretary); Darius Campbell (NEAFC Secretary); Ellen Kenchington (Fisheries and Oceans Canada); Elisabetta B. Morello (GFCM Secretariat); and Aurora Nastasi (GFCM Secretariat).

Contributors

_____

Brynhildur Benediktsdóttir: NAFO Executive Secretary (bbenediktsdottir@nafo.int)

Darius Campbell: NEAFC Secretary (darius@neafc.org)

Ellen Kenchington: Fisheries and Oceans Canada (Ellen.Kenchington@dfo-mpo.gc.ca)

Elisabetta B. Morello: GFCM Secretariat (Elisabetta.Morello@fao.org)

Aurora Nastasi: GFCM Secretariat (Aurora.Nastasi@fao.org).

List of Tables

---

[1] Table 3.1 was developed for a CBD-FAO-FEG meeting on regional capacity building for fishery-OECMs, 17-19 May 2023.

List of Figures

List of Boxes

Abbreviations

ABFM	Area-based fishery management
ABMT	Area-based management tool
BACI	Before-After-Control-Impact
BIM	Biodiversity Impact Mitigation
CBD	Convention on Biological Diversity
CCFAM	Canadian Council of Fisheries and Aquaculture Ministers
CCRF	Code of Conduct for Responsible Fisheries
CITES	Convention on International Trade in Endangered Species of Wild Fauna and Flora
COP	Conference of the Parties to the CBD
EAF	Ecosystem Approach to Fisheries
EBSA	Ecologically or biologically significant marine area
EEZ	Exclusive economic zone
EFH	Essential fish habitat
EFS	Ecosystem functions and services
ES	Ecosystem services
FAO	Food and Agriculture Organization of the United Nations
ICAM	Integrated coastal area management
ILK	Indigenous and local knowledge
IP	Indigenous Peoples
IUCN	International Union for Conservation of Nature
ICES	International Council for the Exploration of the Sea
FEG	Fisheries Expert Group of the IUCN Commission on Ecosystem Management
LME	Large marine ecosystem
MCS	Monitoring, control and surveillance
MER	Monitoring, evaluation and reporting
MPA	Marine protected area
MSP	Marine spatial planning
NTZ	No-take zone
OECM	Other effective area-based conservation measure
OSPAR	Convention for the Protection of the Marine Environment of the Northeast Atlantic

PA	Protected area
PET	Protected, endangered or threatened (species)
RFMO/A	Regional fishery management organization/arrangement
RSC	Regional Seas Convention
PSR	Pressure/state/response framework
SAI	Significant adverse impact
SBSTTA	Subsidiary Body on Scientific, Technical and Technological Advice
SCBD	Secretariat of the Convention on Biological Diversity
SDG	UN Sustainable Development Goal
SSF	Small-scale fishery
TURF	Territorial Use Rights in Fisheries
UN	United Nations
UNCLOS	United Nations Convention on the Law of the Sea
UNGA	United Nations General Assembly
UNEP	United Nations Environment Programme
UNFSA	United Nations Fish Stock Agreement
VME	Vulnerable marine ecosystem
WCC	World Conservation Congress
WCMC	World Conservation Monitoring Centre (UNEP)
WCPA	World Commission on Protected Areas
WDPA	World Database on Protected Areas

1 Introduction

Abstract

Since the 1990s, fisheries management has had to increase its attention on broader biodiversity features and the role of area-based management tools (ABMTs) to reduce its collateral impact and enhance conservation. The Other Effective Area-based Conservation Measures (OECMs) emerged in 2010 in the CBD Aichi Target 11. Their definition was adopted in the 2018 CBD Decision 14/8, together with Principles, and Criteria, and their role in global conservation coverage targets was confirmed by the Global Biodiversity Framework in 2022. Decision 14/8 called for economic sectors including fisheries to identify OECMs in their areas of competence. As a response, efforts to promote fishery-OECMs, initiated by the IUCN Fisheries Expert Group in collaboration with CBD and FAO staff, in collaboration with ICES and a few RFMOs, led to the first set of identifications in NEAFC and NAFO in 2025.

Through the 1990s and 2000s, it became clearer in managing fisheries as well as other sectors involving sustainable use (such as forestry and hunting) that (i) biodiversity conservation was intrinsic for any sustainable use (IPBES 2022a: 2.2.2), and (ii) spatial and non-spatial measures need to be combined to really live in *harmony with nature.*[1] Area-based fisheries management measures (ABFMs) have been used for centuries in both traditional and modern fisheries management. They have been defined as follows.

> ABFMs are formally established, spatially-defined, fishery management and/or conservation measures, implemented to achieve one or more intended fishery outcomes, commonly related to sustainable use of the fishery. However, they can also often include protection of, or reduction of impact on, biodiversity, habitats, or ecosystem structure and function. (CBD, 2018)

A detailed inventory of ABFMs can be found in Rice *et al.* (2018) and Garcia *et al.* (2024). ABFMs conventionally aim solely to optimize fish harvesting but, particularly since the United Nations General Assembly (UNGA) Resolution 61/105 on Vulnerable Marine Ecosystems (VMEs) (United Nations, 2007a), many aim also to protect habitats and biodiversity. From this broader perspective, ABFMs link closely to the recent emergence of Other Effective Area-based Conservation Measures (OECMs).

In 2010, reference to OECMs appeared for the first time during the extended negotiations of the 10th Conference of the Parties to the Convention on Biological Diversity (CBD COP 10), in Nagoya, Japan. The Conference adopted a Strategic Plan for Biological Diversity 2011–2020 containing 20 targets (referred to as Aichi Targets) to be reached, in most cases, by 2020. The general focus of each Target had been set before the COP began, with Target 11 to set the goal for area-based conservation efforts. Target 11 proved to be among the most challenging for reaching consensus, with many NGOs and some Parties determined to have it exclusively count fully and legally protected areas, where other Parties and some organizations were comparably determined to allow a range of other spatial tools to be included in reaching the Target coverage. During the COP a Contract Group charged to find consensus on Target 11 proposed several alternative wordings to the Plenary but none received consensus. Over the second week of the COP, the Contact Group slowly negotiated a phrase to capture the properties necessary for areas other than Protected Areas to be included under Target 11, inclusive enough to get consensus yet stringent enough to satisfy concerns on effectiveness. An acceptable phrasing was finally reached before the COP concluded, and the Plenary adopted Aichi Target 11 stating:

> by 2020, at least ... 10 per cent of coastal and marine areas, especially areas of particular importance for biodiversity and ecosystem services, are conserved through effectively and equitably managed, ecologically representative and well-connected systems of protected areas (PAs) and *other effective area-based conservation measures (OECMs)*, and integrated into the wider landscapes and seascapes[2] (brackets and emphasis added).

In approving this Target, the CBD COP established a new conservation concept, now systematically referred to as 'OECM' although the CBD Parties never adopted any abbreviation. It is crucial to note that the term 'OECM' does not refer to a new category of area-based management tools (ABMTs) but to a collection of various conservation areas to which an OECM label and status are granted because they meet the relevant criteria. This allows OECMs to be counted, together with PAs against global conservation coverage targets.[3]

After 2010, decisions taken at each subsequent CBD COP began to actively develop bounds and criteria for OECMs, as follows.

- In 2012, Decision XI/18 on Marine and Coastal Biodiversity agreed to organize training workshops, in order to increase the capacity of Parties to use marine spatial planning as a tool to enhance existing efforts in integrated marine and coastal area management, including marine protected area networks and other area-based management efforts, and other marine biodiversity conservation and sustainable use practices (§C1g).
- In 2012, Decision XI/24 on Protected Areas agreed, consistent with national circumstances, to undertake major efforts, with appropriate support, to achieve all elements of Aichi Biodiversity Target 11, inclusive of diversified types of governance for protected areas and other effective area-based conservation measures (§1b).
- In 2016, Decision XIII/2 on Progress towards the achievement of Aichi Biodiversity Targets 11 and 12 in which COP XIII agreed to:

  ○ pursue efforts to identify and explore options to protect areas of particular importance for biodiversity and ecosystem services, in establishing new and/or expanding existing protected areas, or taking other effective area-based conservation measures, to give due consideration to areas that: (i) improve ecological representativeness; (ii) increase connectivity; (iii) promote the integration into the wider landscape and seascape; (iv) protect the habitats; (v) expand the coverage of areas important for biodiversity and ecosystem services; (vi) are identified as centres of origin or centres of genetic diversity; (vii) are managed under collective action by Indigenous Peoples and local communities (§5b);

  ○ undertake more systematic assessments of management effectiveness and biodiversity outcomes of protected areas, and where possible, other effective area-based conservation measures (§5c);

  ○ invite Parties, other governments, relevant partners, regional agencies, bilateral and multilateral funding agencies to undertake a review of experiences on: (i) protected areas and other effective area-based conservation measures; (ii) additional measures to enhance integration of protected areas and other effective area-based conservation measures into the wider land- and seascapes; and (iii) mainstreaming of protected areas and other effective area-based conservation measures across sectors (§9);

  ○ organize a technical expert workshop or workshops to provide scientific and technical advice on definition, management approaches and identification of other effective area-based conservation measures and their role in achieving Aichi Biodiversity Target 11 (§10b); and

  ○ invite the Global Environment Facility and its implementing agencies to facilitate the alignment of the development and implementation of protected area and other effective area-based conservation measures.

These Decisions indicate the efforts the CBD invested in trying to get a broader interpretation of conservation, involving both protected areas and OECMs. In 2018, building on progress made at preceding COPs, the CBD COP XIV Decision 14/8 addressed all the elements noted above, adopting the OECM definition (see Section 2.2). It also provides identification criteria, common principles and guides for equitable governance, as foundations for effective implementation (Decision 14/8, Annex III). More details are provided in Chapter 3.

Most importantly, in 2022, the Kunming-Montreal Global Biodiversity Framework (GBF) adopted a Target 3 which updated Aichi Target 11 and, *de facto*, the bumper date for its achievement, but raising the 'stake' to a 30% global coverage of marine protected areas (MPAs) and OECMs, confirming their potential role in key economic sectors[4] and stressing the role of Indigenous Peoples and local communities in the process.

Clearly willing to foster mainstreaming of OECMs in economic sectors, Decision 14/8 (§9) '… invites the International Union for Conservation of Nature, the Food and Agriculture Organization of the United Nations, and other expert bodies to continue to assist Parties in identifying other effective area-based conservation measures and in applying the scientific and technical advice'. The Decision (§12) also '… urges Parties to facilitate mainstreaming of protected areas and other effective area-based conservation measures into key sectors, such as, inter alia, agriculture, *fisheries*, forestry, mining, energy, tourism and transportation'. In addition, Annex I of Decision 14/8 calls for integration of protected areas and OECMs into wider landscapes/seascapes and biodiversity mainstreaming across sectors.

As a first response to this invitation, a global 'Expert Meeting on Other Effective Area-Based Conservation Measures in the Marine Capture Fishery Sector' was organized by FAO, the IUCN-CEM Fisheries Expert Group (FEG) and the European Bureau for Conservation and Development (EBCD), in collaboration with the CBD Secretariat (7–10 May 2019, Rome, Italy) (FAO, 2019a). The purpose of the expert meeting was to compile a broad range of expert advice on the identification and establishment of OECMs in the marine capture fishery sector, on the basis of Decision 14/8. The expert meeting considered a range of topics: (i) the rationale for producing guidance for OECMs in the marine capture fishery sector; (ii) definition of an OECM; (iii) guiding principles and common characteristics; (iv) criteria for identification and evaluation; (v) key concepts and cross-cutting issues in a fisheries context; (vi) evaluating areas for inclusion in OECM reporting and management; (vii) monitoring, evaluation and reporting; (viii) re-evaluation of the OECM; and (ix) selected governance issues. Discussions were supported by a background document addressing these issues (Garcia *et al.*, 2019).

A conclusion of the above meeting was that regional meetings would be necessary to pursue the global reflection in different social and economic contexts, differing in terms of data, scientific assessment, management and related capacity-building needs. Consequently, a first regional Workshop[5] on 'Testing OECM Practices and Strategies (WKTOPS)' was organized by the International Council for the Exploration of the Sea (ICES) and the IUCN-CEM Fisheries Expert Group (FEG) (15–24 March 2021) (ICES, 2021). The workshop examined the elements of guidance available in a background document prepared by FEG for the purpose and focused on the actions that might be considered to implement Decision 14/8 (Garcia *et al.*, 2021), in the context of six case studies. Subsequently, other regional meetings with some focus on fisheries-OECMs have been organised, including in the east and south Mediterranean (IUCN-WCPA, 2020), the Baltic Sea (HELCOM, 2022), the wider Caribbean (FAO and SCBD, 2023), Jamaica (CBD, 2023a), and North and West Africa (CBD, in February 2025[6]).

In addition, many governments are now involved in mainstreaming OECMs in their conservation agenda. A number of countries including Canada, Colombia, Morocco, Algeria, Philippines, Jamaica and the European community have started identifying OECMs. The government of Canada, through its Department of Fisheries and Oceans, has developed a specific guidance for fisheries (Government of Canada, 2022).

Many of the OECMs identified or being considered in the marine domain are in coastal areas and seem mostly to apply to areas of interest for conservation within or around which fisheries and tourism may operate. Progress on offshore OECMs has been slower, with many Parties reluctant to act until the BBNJ Treaty is ratified and establishes additional governance processes for conservation of biodiversity in Areas Beyond National Jurisdiction (ABNJ).

Regional fisheries management organizations (RFMOs) have also started to identify OECMs in the fisheries context, most notably the Northwest Atlantic Fisheries Organization (NAFO) and the North-East Atlantic Fisheries Commission (NEAFC) in relation to vulnerable marine ecosystems (VMEs), and the General Fisheries Commission for the Mediterranean (GFCM) in relation to its Fisheries Restricted Areas (FRAs) (see Boxes 2.1, 4.1 and 4.2 for more detail).

Notes

[1] The centuries-old objective and vital need to live in harmony with nature is still the motto and vision of the Convention on Biological Diversity (CBD) strategic programme on biodiversity conservation towards 2050.

[2] CBD Decision X/2 (2010): www.cbd.int/decision/cop?id=12268. https://www.cbd.int/sp/targets/ (accessed 17 July 2025).

[3] Such as Aichi and GBF Targets, in line with Decision XIV/8, paragraph 12 and its Annex I, and in accordance with the objectives of the CBD Article 1 and related obligations, including those related to *in situ* and *ex situ* conservation (Articles 8 and 9), and sustainable use (Article 10) of biodiversity and its components.

[4] Target 3 aims to 'ensure and enable that by 2030 at least 30 per cent of terrestrial, inland water, and of coastal and marine areas, especially areas of particular importance for biodiversity and ecosystem functions and services, are effectively conserved and managed through ecologically representative, well-connected and equitably governed systems of protected areas and other effective area-based conservation measures, recognizing indigenous and traditional territories where applicable, and integrated into wider landscapes, seascapes and the ocean, while ensuring that any sustainable use, where appropriate in such areas, is fully consistent with conservation outcomes, recognizing and respecting the rights of indigenous peoples and local communities, including over their traditional territories'.

[5] Held virtually because of COVID.

[6] www.cbd.int/doc/notifications/2024/ntf-2024-106-marine-en.pdf (accessed 17 July 2025).

2 The CBD Decision 14/8

Abstract

The chapter reviews the foundations of OECMs provided by the 2010 Aichi Target 11 on global conservation coverage (now superseded by the GBF Target3) and the 2018 CBD Decision 14/8. It examines the implications of Decision 14/8 for fishery-OECMs, in relation to the definition, identification criteria, common principles, equitable governance and effective management. It underlines the flexibility allowed in the implementation of fishery-OECMs, allowing case-by-case adaptation to local conditions but emphasizing that all key elements of the definition should be duly considered, and stressing the strong requirement for positive and sustained long-term biodiversity outcomes, while accounting for locally relevant social and economic values. It also provides background considerations on issues like complementarity, mainstreaming, integration, representativeness and connectivity which are important additional properties of OECMs.

As noted in the previous chapter, Decision 14/8 was adopted by CBD Parties in 2018 to allow a practical and consistent implementation of the 2010 Aichi Target 11, and to clarify the nature, role, identification process and use of OECMs at site, national, network and sectoral levels, in all ecosystems. To see more clearly how this CBD Decision is shaping and will continue to shape the fishery approach to OECMs, details of the nature and content of the Decision are examined below.

2.1 Properties of Conservation Coverage Targets

It is helpful to consider the CBD Decision in the context of the foundational Aichi Target 11, which was adopted in 2010 by CBD COP 10 as part of the 20 Targets contained in the Strategic Plan for Biological Diversity 2011–2020 (referred to as Aichi Targets). Aichi Target 11 introduced OECMs, stating:

> By 2020, at least 17 per cent of terrestrial and inland water, and 10 per cent of coastal and marine areas, especially areas of particular importance for biodiversity and ecosystem services, are conserved through effectively and equitably managed, ecologically representative and well-connected systems of protected areas

and other effective area-based conservation measures [later referred to as OECMs], and integrated into the wider landscapes and seascapes.

The 2010 Strategic Plan and its Aichi Target 11 have now been superseded by the 2022 Global Biodiversity Framework and its GBF Target 3 that aims to:

> ... ensure and enable that by 2030 at least 30 per cent of terrestrial, inland water, and of coastal and marine areas, especially areas of particular importance for biodiversity and ecosystem functions and services, are effectively conserved and managed through ecologically representative, well-connected and equitably governed systems of protected areas and other effective area-based conservation measures, recognizing indigenous and traditional territories where applicable, and integrated into wider landscapes, seascapes and the ocean, while ensuring that any sustainable use, where appropriate in such areas, is fully consistent with conservation outcomes, recognizing and respecting the rights of indigenous peoples and local communities, including over their traditional territories.

GBF Target 3 retains all the MPAs and OECM properties mentioned in Aichi Target 11, that provided the foundation of the OECMs definition, criteria and expected properties. It also increases the target conservation coverage to 30%. In addition, the GBF calls for 'recognizing and respecting the rights of indigenous peoples and local communities, including over their traditional territories', as well as the legitimacy of sustainable use of biodiversity resources in OECMs, as long as they are *fully consistent with conservation outcomes*.

In 2010, the CBD Parties identified these properties against a background of growing concerns expressed over threats to a degrading biodiversity under various stressors, including unsustainable uses, habitat loss and climate change; the variable effectiveness of existing protected areas and the lack of systematic assessment of their performance; the sluggish mainstreaming of biodiversity concerns in economic sectors, including fisheries; the failure to achieve global conservation coverage targets; the reluctance of some traditional communities to register their conservation areas as marine protected areas (MPAs), and particularly as no-take areas; the gaps and slow development in spatial conservation; and the lack of accounting of conservation achieved by other measures than MPAs (Lopoukhine and Ferreira de Souza Dias, 2012:2; Visconti *et al.*, 2019; Hilborn and Sinclair, 2021).

The expression 'other effective area-based conservation measures' was intended to allow numerous conservation areas – other than formal protected areas – regulated for different purposes but demonstrably producing biodiversity conservation benefits, to be considered for inclusion in the Target. The properties required in the Target for the integration of such 'other measures' are (i) their importance for biodiversity and related ecosystem services; (ii) their effective and equitable management, addressing current and future threats; (iii) their representativeness and connectivity within conservation networks; and (iv) their integration in landscapes or seascapes. We will see below that these properties are reflected in Decision 14/8, in the definition of OECMs, the guiding principles and criteria for their identification, and the guidance on their integration and equitable governance.

Aichi Target 11 reflected both the inherent complexity of the conservation challenges being addressed, and the difficulty of building consensus at the COP, including: (i) the need to address the different starting conditions for coverage of terrestrial and marine conservation areas in different parts of the world; (ii) the fact that the 'other areas' should cover a diversity of areas, ranging from terrestrial indigenous lands under traditional agroforestry practices to various marine spatial measures other than formal MPAs, established by conservation institutions or economic sectors; (iii) the fact that tenure conditions are different on land and at sea; and (iv) that the governance structures and processes for fully protected areas in both terrestrial and marine systems would often be substantially different from governance of 'other areas', particularly if managed by economic sectors. On land, a range of forms of land use and property rights exist, whereas in the ocean, property ownership does not exist (except under very special conditions and in the territorial sea), and the States are granted conditional use rights in their exclusive economic zone (EEZ), that they can allocate as they wish. Consequently, different approaches and regulations may be needed on land and in the ocean, and under different social and economic conditions, to achieve the same outcome regarding uses and conservation of ecosystem features.

Nonetheless, for the OECM concept to be applied consistently at national level, across sectors, in the ecosystem and hopefully worldwide, there needed to be a minimum of consistency in the interpretation by different countries of which area-based conservation measures (other than MPAs) may be considered as 'effective area-based conservation measures'. The basis for such consistency is to be found in the definition and criteria contained in Decision 14/8 and is reviewed below.

2.2 Nature and Content of Decision 14/8

Decision 14/8, referred to hereafter as 'the Decision', was adopted by the CBD Conference of the Parties (COP) in 2018. It is an international policy instrument that supports the implementation and interpretation of the Convention and of its previous COP decisions. However, it can be noted that the section with some degree of 'bindingness' in the Decision is its paragraph 2 by which the CBD COP *adopts* formally the following OECM definition.

A geographically defined area other than a Protected Area, which is governed and managed in ways that achieve positive and sustained long-term outcomes for the *in situ* conservation of biodiversity,[1] with associated ecosystem functions and services and where applicable, cultural, spiritual, socio–economic, and other locally relevant values. (Decision 14/8, §2)

This definition is therefore the strongest part of the Decision and its content needs to be carefully kept in mind to remember the spirit of that Decision. All other paragraphs of the Decision core text are drafted more 'softly', *welcoming, encouraging, inviting, urging* State Parties and other governments to take specific actions. In addition, the four annexes contain what is explicitly referred to as *voluntary guidance* or *scientific and technical advice* and they are therefore not binding.

The relative 'softness' of the Decision is reinforced by the short preamble to its Annex III which states that *the guiding principles and common characteristics and criteria for identification of OECMs are applicable across all ecosystems currently or potentially important for biodiversity, and should be applied in a flexible way and on a case-by-case basis*, confirming that a significant amount of implementation details is left to the appreciation of States and other legitimate authorities. However, this implementation flexibility must be consistent with the binding definition adopted in Paragraph 2 of the Decision (see Section 2.3) which clearly indicates that OECMs are *governed and managed in ways that achieve positive and sustained long-term outcomes*, indicating that the factor that determines the OECM status is the outcomes or benefits it generates through governance and management (i.e. what is achieved and sustained), and not any specific features of how the area is governed and managed.

The first two pages of Decision 14/8 contain the formal definition of OECMs, and indicate what action is expected from State Parties, other governments and the CBD Secretariat. The following 17 pages contain four annexes dealing respectively with: (i) the integration of protected areas and OECMs in landscapes- and seascapes, and their mainstreaming across economic sectors (Annex I); (ii) properties of effective governance models for protected areas (Annex

II); (iii) scientific and technical advice on OECMs (Annex III); and (iv) considerations in achieving Target 11 in marine and coastal areas, including lessons learned (Annex IV).

Altogether, Decision 14/8 is an important international policy instrument, reflecting a substantial political commitment to mainstream OECMs across ecosystems and sectors. All parts of the Decision ought to be considered for its faithful implementation, but its spirit (and explicit wording) leave enough flexibility to State Parties, in co-operation with the stakeholders, to adapt the process to local conditions. This does not mean, however, that anything can be done under the OECM status. The Decision has been adopted in good faith and it reflects a political commitment of all CBD Parties, even though no formal mechanism exists to ensure compliance. This definition contains all the properties expected from OECMs, leading to and further elaborated in all the advice and voluntary guidance contained in the Decision. So, despite the flexibility allowed explicitly by the Decision, the key elements composing the definition should (or ought to)[2] be adhered to.

In this book, we nonetheless often refer to things that '*need* to be done' but are not explicitly in the definition, and this requires a clarification. If a State decides to identify OECMs, the OECM definition should be observed. The implementation flexibility remains but there are scientific and technical 'needs', not imposed by the CBD Decision but implied by scientific and technical logical consequences that the Decision only suggests or does not mention. For example, the definition *requires* that to be granted the OECM status, an area *should* be *effectively managed*. This implies the existence of some management capacity. For fishery-OECMs, this would be reflected in objectives, regulatory measures and ways to show that objectives are *effectively* being reached. For OECMs identified and managed by Indigenous Peoples or local communities, however, 'effectively managed' could be achieved without the types of explicit management objectives and regulations and ecological performance metrics commonly used by sectoral managers. These sectoral elements are not mentioned in the definition, but they appear in the Decision Annex II as *voluntary guidance* and in the Principles and Criteria for Identification (Decision Annex III) as '*scientific* and technical advice', in the softer part of the Decision. Similarly, Decision 14/8 (Annex III, B) lists, for each criteria, 'elements of evidence' that might be considered during the assessments. These are neither exhaustive nor mandatory, but collecting and providing evidence is a scientific necessity to justify the granting of the OECM status and its maintenance in the long term. This is important to limit the risk of 'green washing' in the implementation of the GBF Target 3. The lack or weakness of such 'evidence' only weakens the OECM identification case, unless it relates to an OECM property considered as 'fatal'. Moreover, consistent with the term 'sustained' in the definition, if the State intends to maintain the OECM status in the long term, it needs to check and report on its performance regularly or occasionally, with whatever means it has at disposal.

On the same issue, but from a different angle, this book aims to illustrate what an effective OECM implementation cycle could look like in order to faithfully implement Decision 14/8 and really contribute to biodiversity conservation. The phases and steps proposed in the implementation cycle (Chapter 3) and identification phase (Chapter 5) are not obligations. Effective alternatives are possible. Some steps might be inverted or some combined or undertaken in parallel, depending on context. Therefore, the phases and steps of the cycle described in the book are not mandatory. However, if fully implemented, the cycle as described here is *sufficient* to achieve the intent of OECMs. If phases or steps are simply by-passed, rather than done in a different way, difficulties might emerge and the OECM status might appear questionable or inappropriate. Consequently, many of the elements we refer as 'needs' or those accompanied by the verb 'should' are scientific and technical necessities and not mandatory.

In the following sections, we will consider the definition of OECMs and the scientific and technical advice contained in Annex III of the Decision.

2.3 Understanding OECMs

'Other Effective Area-based Conservation Measures' are conveniently referred to by their abbreviation OECM, and are both conservation 'areas' and conservation 'measures'. However, as often stressed across this book, OECMs are primarily 'areas', i.e. special places where technical regulations are applied to obtain long-term positive biodiversity outcomes (or benefits) and, most importantly, to which the OECM status is granted. The spatial dimensions of OECMs, such as geographical location, boundaries, depth and spatial subdivisions, are therefore fundamental.

The general definition of OECMs given above (see Section 2.2) is the strongest part of Decision 14/8. It is applicable *across all ecosystems* and *a priori* to all sectors operating in these ecosystems and with potential to generate significant biodiversity benefits through their spatial measures, whether existing, improved or created *ex nihilo*. Consequently, as long as the OECM criteria are met, the OECM status may be granted to large multiple-use conserved areas in which some fisheries operate (like community-managed areas) as well as to smaller areas managed by fisheries within which fishing activities are constrained.

The OECM definition contains the key properties of the area regarding (i) the geographical definition, (ii) legal status, (iii) governance and management, and (iv) achievements, accounting for other locally relevant values (e.g. of a socioeconomic or cultural nature). These properties are reflected in the Identification Criteria, the Common Principles and the voluntary guidance provided by Decision 14/8. The implication is that although State parties may flexibly implement the Decision, these four properties represent the policy foundation of OECMs and the spirit of the Decision.

The adoption of the OECM definition is a recent policy development but it is possible to trace the legal basis for OECMs to the Convention itself (Johnson et al., 2024). CBD Article 8(a) on *in situ* conservation mandates Parties to 'as far as possible and as appropriate'... 'establish a system of protected areas *or areas where special measures need to be taken to conserve biological diversity*' (emphasis added). *In situ* conservation of biodiversity is one of the obligations of Parties to the CBD with a view to achieving two objectives of the Convention regarding the conservation of biodiversity and the sustainable use of its components (CBD, Art. 1). Therefore, the identification of OECMs in accordance with the definition adopted in CBD Decision 14/8 can support the implementation of not only GBF Target 3 but also Article 8(a) of the Convention itself.

At site level, the identification of an OECM, its effectiveness and eventual comparison with other OECMs rest on two sets of features: (i) the site 'area' with its specific location in the ecosystem, its geographically localized boundaries and eventual subdivisions (zones), its extension (needed for the coverage target), the biodiversity values it hosts and which are protected; and (ii) the regulations applied in and possibly around the 'area', such as restrictions of access and activities allowed or excluded in the 'area' to reach the assigned objectives.

2.3.1 Comparison of PAs and OECMs properties

Protected areas and OECMs have common properties and key differences. These properties emerge from their specific definition, complemented by the specifications adopted for Aichi Target 11 and GBF target 3 (Table 2.1). In the fisheries context, fishery-OECMs are special area-based fisheries management measures (ABFMs) to which an OECM status has been granted. PAs, ABFMs and OECMs are area-based management tools (ABMTs), a term covering over 80 spatial management instruments (Garcia and Rice, 2024). Some definitions have been provided before but are repeated below for easier access and comparison.

- *ABMT*: a tool, including marine protected areas, used to manage activities or sectors within a geographically defined area to achieve specific conservation and sustainable use objectives. These tools are designed to enhance the conservation and sustainable use of marine biodiversity within a designated area (BBNJ Agreement, Part 1, Article 1).[3]
- *PA*: 'a geographically defined area, which is designated or regulated and managed to achieve specific conservation objectives' (United Nations, 1992b: Article 2). 'A clearly defined geographical space,

recognised, dedicated, and managed, through legal or other effective means, to achieve the long-term conservation of nature with associated ecosystem services and cultural values' (IUCN in Dudley, 2008).
- *ABFM*: 'a formally established, spatially-defined, fishery management and/or conservation measure, implemented to achieve one or more intended fishery outcomes, commonly related to sustainable use of the fishery. However, it can also often include protection of, or reduction of impact on, biodiversity, habitats, or ecosystem structure and function' (CBD, 2018).
- *OECM*: 'a geographically defined area other than a Protected Area, which is governed and managed in ways that achieve positive and sustained long-term outcomes for the *in situ* conservation of biodiversity, with associated ecosystem functions and services and where applicable, cultural, spiritual, socio-economic, and other locally relevant values' (CBD, 2018: Decision 14/8, §2).

Both the CBD and IUCN consider their definitions of PAs as equivalent. However, the IUCN definition, adopted many years later than the CBD one, adds the need for effective management and the requirement for long-term conservation benefits. It also specifies that the conservation target is broad ('nature'), and introduces the notions of ecosystem services and cultural values. These specifications have also been identified as important for OECMs in Decision 14/8, except that the conservation target of OECMs is '*in situ* biodiversity outcomes' and not 'nature' which is usually understood as a broader, all-encompassing concept embracing geodiversity, landform and broader natural values (Dudley, 2008).

The OECM properties referred to in their definition derive from fundamental properties of effective conservation areas specified in Aichi Target 11 and GBF Target 3, and specific guidance contained in the Decision 14/8 such as delivering long-term positive biodiversity outcome; ecological representativity; complementarity and connectivity with MPAs within conservation networks; sustained effective management; equitable governance; taking into account ecosystem services and other locally

Table 2.1. Properties indicated in the definitions of protected areas, Aichi Target 11, GBF Target 3 and OECMs of relevance in marine areas.

Properties		Protected area definitions		OECMs[a]	Aichi Target 11[b]	GBF Target 3
		CBD	IUCN			
Objectives/outcomes	Coverage				At least 10%	At least 30%
	Biodiversity	Conservation	Nature	*In situ* biodiversity	Biodiversity	Biodiversity
	Ecosystem	n.a.	Services	Functions and services	Functions and services	Functions and services
	Other values	n.a.	Cultural values	Cultural, spiritual and socioeconomic	n.a.	IP and LC values
	Governance	n.a.	n.a.	Governed: legitimate, diverse, inclusive, equitable	n.a.	Effective, equitableRespects IP and LC rights and territories
	Management	Regulated Managed	Managed Effective	Managed: achieve, effective, equitable	EffectiveEquitable	Effective Equitable
	Legal status	Designated	Legal, dedicated, recognized	Not a protected area, identified	n.a.	Ensured, enabled
	Geolocalization	Geographically defined	Geographically space	Geographically defined	n.a.	n.a.
	Benefits	n.a.	Long term	Positive, sustained, long term	n.a.	Sustainable use, consistent with conservation outcomes
	Ecological representation	n.a.[c]	n.a.	Representative	Representative	Representative
	Connected	n.a.	n.a.	Networks, systems	Well-connected systems of PAs and OECMs	Well-connected systems of PAs and OECMs
	Integration	n.a.	n.a.	Seascapes; cross-sectoral	Seascapes	Seascapes, ocean

n.a., not available, i.e. that property is not mentioned.
IP and LC, Indigenous Peoples and local communities; PA, protected area.
[a]Additional properties referred to in the voluntary guidance are in italics.
[b]Aichi Target 3 has been superseded by GBF Target 3.
[c]But ecological representativity is referred to in CBD Decision IX/20, Annexes II and III.

relevant values. These properties apply *de facto* also to MPAs as effective area-based conservation measures.

Differences and similarities between PAs and OECMs can be inferred from their definitions, the properties required in Aichi Target 11 and GBF Target 3, as well as from Decision 14/8 and its guidance, and are addressed, for example, in UNEP-WCMC (2019),[4] IUCN-WCPA (2019) and Garcia *et al.* (2021). The properties on protected areas and OECMs as inferred from their definition, as well as the properties specified in Aichi Target 11 and GBF Target 3, are given in Table 2.1.

The similarities are logical because according to Aichi Target 11 and GBF Target 3, both MPAs and OECMs are 'effective area-based conservation measures' the properties of which apply to both, complementing their definitions.

Similarities between MPAs and OECMs include: (i) the geographically defined localization; (ii) the conservation purpose (but see below); (iii) the need for management effectiveness; and (iv) the role of a participative and equitable governance, particularly the explicit acknowledgement of the rights and legitimacy of Indigenous Peoples and local communities and their traditional territories. In addition, both the CBD OECM guidance and GBF Target 3 stress that both PAs and OECMs – which jointly contribute to global conservation coverage targets – contribute to improve ecological representativeness and connectivity in conservation networks and integration across seascapes.

The differences include the following: (i) the OECM status cannot be granted to areas already designated as MPAs (in order to avoid double-counting towards the global conservation coverage target); (ii) PAs are usually formally 'designated' by a central or community environmental authority, while OECMs may be 'identified or recognized' by a sectoral or environmental authority, preferably working in co-operation; (iii) PAs aim to sustain or improve 'nature as a whole' while OECMs are expected to generate specific positive long-term outcomes; (iv) PAs have a broad conservation goal as the primary objective, while OECMs will usually have their sectoral purpose (such as sustainable use of fishery resources) as the primary objective, with the production of positive biodiversity outcomes as either a primary

or secondary objective or as identified albeit sometimes unintended co-benefits.[5] In any case, the sustainable use regime should be consistent with the biodiversity-related objectives; (v) PAs may restrict all impacting activities within their boundaries (e.g. in strict reserves) but may allow a range of activities (e.g. in miltiple-use PAs). By contrast, sectoral OECMs may only control the sectoral activities,[6] but may be estblished under cross-sectoral collaboration; (vi) PAs tend to be formally considered as 'protected' by their status, irrespective of their effectiveness,[7] while OECMs must demonstrate their effectiveness or intention to be effective, and their OECM status may be cancelled if they fail to perform as planned.

While the differences are clear, some ABFMs identified as fishery-OECMs may also be areas that meet the PA definition but the legitimate authority prefers to consider them as OECMs.

OECMs and fishery-OECMs in particular are not a new and separate category of spatial fishery measures (of which numerous examples are given in Rice *et al.*, 2022 and Garcia *et al.*, 2024) but rather a specific CBD international status or label,[8] granted by legitimate authorities to existing or new area-based measures if they meet the OECM criteria, independently of their primary objective. Considering the importance and sensitivity of the local names of a specific place in various legal or regulatory documents, when granted OECM status, such measures may maintain their local ABFM name[9] and original category[10] but benefit from an international recognition of the biodiversity benefits they generate or are expected to generate.

2.3.2 Application of decision 14/8 to fishery-OECMs

The term 'fishery-OECM' or 'fisheries-OECMs' was introduced in 2020 in two related background documents prepared by the IUCN Fisheries Expert Group (Garcia *et al.*, 2020a, b). It was taken up in ICES (2021) and finally in the FAO Handbook on OECMs in fisheries (FAO, 2022a). The term was introduced because OECMs can be established in different sectors or directly for biodiversity protection and hence are likely to differ somewhat in their implementation details, even though they all meet the same

CBD definition and criteria. Paraphrasing the CBD general definition given above (see Section 2.3), a fishery-OECM can be defined as:

> a geographically defined area other than a Protected Area, which is governed and managed by a mandated fisheries authority in ways that achieve positive and sustained long-term outcomes for the *in situ* conservation of biodiversity, with associated ecosystem functions and services and where applicable, cultural, spiritual, socioeconomic and other locally relevant values, along with any intended fishery outcomes.

Other similar definitions have been proposed (e.g. in FAO, 2022a) but this one is particularly convenient because: (i) it illustrates that a fishery-OECM can be any area faithfully meeting the OECM definition and criteria, identified by a legitimate fishery authority within its arsenal of existing or newly designed spatially based fishery measures; and (ii) it acknowledges that fishery-OECMs often have two broad objectives: to ensure the conventional sustainability of the fishery and to generate long-term benefits, particularly for the biodiversity features of concern. The relative priority among them is less important than the production of both the expected long-term stock and biodiversity benefits.

Many general OECM properties mentioned in Table 2.1 need to be carefully considered when identifying and using OECMs in the marine fishery sector, using a marine and fishery-specific interpretation 'lens', including the following.

- Its geographical definition may need to include depth (e.g as in NEAFC OECMs) in the tridimensional oceanic systems and may be particularly complex in very large, fluid and mobile pelagic systems.
- Numerous closed areas are implemented in fisheries management for a variety of purposes. Case-by-case evaluation for sustained biodiversity co-benefits will be necessary, because effectiveness of specific types of spatial measures may differ depending on the ecosystem and fishery in which they are used (e.g. pelagic vs benthic systems; community-based vs top-down managed fisheries, etc.) and the way in which they are enforced.

- In the ocean, contrary to what happens on land, the only legitimate authority internationally recognized in the United Convention on the Law of the Sea (UNCLOS) is the State or any other authority mandated or recognized by the State, individually or collectively (e.g. in RFMOs). Private property is practically non-existent but complex situations may exist in hybrid areas in the coastal domain.
- Today, some form of management exists in most fisheries with varying degrees of success, but it is crucially missing in very numerous small-scale fisheries in coastal communities, some of which might be effectively managed under traditional, shared or devolved management.
- Effectiveness of fisheries management is logically expected by all stakeholders (except those operating in illegal, unreported and unregulated (IUU) fisheries), but many factors affect how well it is achieved: (i) the means available for management; (ii) the nature of the fish harvesting (gears and practices); and (iii) externalities not under the control of either managers or harvesters, such as oceanographic anomalies. In OECMs effectiveness in delivering positive biodiversity outcomes becomes an explicit requirement to be demonstrated and recurrently confirmed (see Section 7.4).
- Similarly, equity and equitable participation have been part of good governance principles of fisheries for decades, with varying degrees of acknowledgement and implementation, but they are a necessary property for OECMs including those managed by Indigenous Peoples (Anaya, 1996) and other traditional and modern local communities.
- Experience shows that some closed areas have already been in place for a long time, and most have been maintained over time, even when their duration was not initially specified. However, for closures to be considered OECMs, evidence of intent to maintain the closure in the future is necessary to assure that the long-term positive biodiversity outcomes will be maintained.
- Ecosystem functions and services of direct interest to fisheries include provision of

food and provision of livelihoods, which are usually recorded in managed fisheries. Less is usually known about other services provided by the areas supporting the fisheries, particularly services supporting ecosystem functioning. Credible effort to evaluate how fisheries may affect these other services is needed, and often can best be pursued in close co-operation with the biodiversity conservation authorities.

- The 'other locally relevant values' of economic, social and cultural nature have rarely been transparently taken into account in past management of larger-scale fisheries. However, they have been important sometimes for decades in some small-scale fisheries, even if not explicitly quantified, and are gaining prominence in fisheries of many scales (IPBES, 2022b). In the context of OECMs, there may not be pressure to quantify such values in all candidate areas, but even anecdotal evidence that locally relevant values are being transgressed is likely to be a negative factor in determining OECM status.

The above list of fisheries-related issues is probably incomplete but underlines both the general suitability of the CBD criteria and principles for fisheries and the need for their fishery-specific interpretation (see Chapter 3 for a more detailed treatment of implementation issues).

Decision 14/8 recognizes explicitly the potential role of sectoral ABMTs and specifically that of ABFMs as potential OECMs, and defines them as: 'Formally established, spatially defined fishery management and/or conservation measures, implemented to achieve one or more intended fishery outcomes' (CBD 2018, Annex IV, B,2,c). Such intended fishery outcomes are commonly primarily related to sustainable use of the target species of the fishery, such as the protection of their vulnerable life-stages or essential habitats, or to allocation of space and resources among fishing communities or subsectors. However, increasingly, the intended outcomes include protection or reduction of impact on biodiversity components, habitats or ecosystem structure and function, as with areas established to protect vulnerable marine ecosystems (VMEs) or to reduce or exclude small-mesh

fisheries within the foraging range of seabird colonies.

New ABFMs may be created to be fishery-OECMs from the outset but, to date, most discussion about fishery-OECMs has concentrated on (i) identifying OECMs among existing ABFMs that meet the definition and criteria, producing the expected benefits or co-benefits; or (ii) considering situations in which the characteristics of existing ABFMs may be cost-effectively modified to meet the requirements.

Such characteristics have three main dimensions: (i) *time*, as the ABFM may be permanent, temporary, seasonal or real time; (ii) *space*, as ABFMs are usually precisely geo-localized but may also be mobile and dynamic, or redefined or moved as required, for example by oceanographic oscillations or climate change; (iii) *gears and practices* which are regulated within the area to obtain the expected benefit for the target stocks or non-target resources and habitat (Rice and Garcia, 2018). These characteristics can facilitate or impede obtaining the fishery-OECM status, and may sometimes be adjusted to increase the suitability of the area as an OECM.

ABFMs may be considered as fishery-OECMs if they fulfil the requirements contained in Decision 14/8 regarding in particular biodiversity benefits. One implication is that the current or expected biodiversity benefits need to be specified in the Fisheries Management Plans (FMP) both to guide action and performance assessment and provide evidence that the benefits can be expected in the long term (as long as the FMP is in place). Fisheries management agencies are increasingly specifying the objectives of their management plans (Mardle *et al.*, 2004; Hilborn, 2007) but the practice is far from universal. Moreover, in such plans, specific objectives are usually set for the fishery as a whole and not tied to the individual spatial or non-spatial measures. Consequently, the performance of the fishery can be assessed, but the performance of the single management regulation operating in concert rarely can be. Consequently, the ABFM's objectives alone are an incomplete guide to determine which ones could be considered as OECMs, without a careful case-by-case evaluation (Rice *et al.*, 2018).

The primary objectives in an FMP are strongly interconnected and usually mainly focused on conservation of the target resources, sometimes with consideration of the protection of habitats essential for their reproduction and productivity, and the social and economic sustainability of the fishery. In recent decades, however, this conventional idea of fisheries sustainability has been progressively broadened to build in a more ecosystemic approach (see below) that integrates additional concerns regarding by-catch (initially as a waste of biological and economic resources and a threat to fishing operations in the case of protected species), vulnerable habitats deemed essential for the target species' life cycle and productivity, environment- and climate-driven changes on the ecosystem, and multispecies interactions.

While the existence of an ABFM continues to be primarily justified by its performance in relation to conventional fishery sustainability, the granting of OECM status depends on its performance in relation to a broader sustainability concept integrating the concern for biodiversity beyond target resources. At least since the United Nations Convention on Environment and Development (UNCED) in 1992, and with the adoption of the FAO Code of Conduct for Responsible Fisheries (CCRF) (FAO, 1995) and of the Ecosystem Approach to Fisheries (EAF) (FAO, 2003), the broader sustainability concept is being increasingly understood as including achievement of conservation objectives regarding biodiversity, ecosystem structure (including species composition and trophic chains), functions and services (including food provision), as well as paying greater attention to environmental matters, including climate change (Barange et al., 2018). This broader concept of 'sustainability' was captured in the IPBES Sustainable Use Assessment (IPBES, 2022a), and endorsed by consensus by the 146 members of IPBES. In reviewing current literature and policies in fishing, hunting, forestry, gathering and tourism, the assessment concluded that, in practice, 'sustainable use' has at least seven 'dimensions': (i) status of the target species being used; (ii) status of other species and populations affected by the harvest directly (e.g. fisheries by-catches) or indirectly (e.g.because of links with the target species through the trophic chain); (iii) impacts of harvesting technologies on habitats; (iv) economic performance; (v) amount of high-quality employment; (vi) livelihoods of dependent communities and cultures; and (vii) equity in access to uses and distribution of costs and benefits. Priories among the dimensions can vary widely case by case. For example, local community-based fisheries might not be expected to produce substantial new employment, but poor performance on any dimension (e.g. fishing crews working in unsafe conditions) makes the use unsustainable.

2.3.3 Guiding principles and common characteristics

In Annex III of Decision 14/8, CBD COP 14 proposed 13 guiding principles listing the common characteristics expected from OECMs to be understood and considered when implementing OECMs in the fisheries context. In the following sections, the guiding principles are examined in the order in which they appear in Decision 14/8 but have been divided in two sets: (i) those specifying the role and expected outcomes of OECMs; and (ii) those referring to governance of OECMs. For easier cross-referencing across this book, the 13 principles are identified in Table 2.2

Table 2.2. Guiding principles for OECM identification.

Roles and expected outcomes	Governance
Significant biodiversity value: current or intended Important complementary role Dual role: sustainability and conservation Comparable importance Demonstrated outcomes on biodiversity and threats Representative and connected to MPA systems	Consultation: stakeholders, rights- holders Legitimate authority, sustained outcomes Indigenous People and local communities Cultural and spiritual values Diverse governance systems and actors, incentives, empowerment Best available information: science, traditional and local knowledge Transparency and performance evaluation

by a short-cut title of our own to provide a first synoptic view of the principles, and a letter, from (a) to (m), to facilitate cross-referencing to principles across the book. The full text of the principles will be found in sections 2.3.4 and 2.3.5.

A pragmatic additional 'principle', not explicit in this part of the Decision 14/8 but implicit in the concepts of 'flexible implementation' and 'case-by-case' approach, is that no matter how detailed and precise the 'requirements' for OECMs look like, including in this book, countries are only expected to use the best 'available knowledge available' and the best assessment and competence management they can afford. It would indeed be counterproductive if the exigence of the absolute rigid 'best' would become the enemy of 'the better'. The implication of this soft approach is that the OECM identification and maintenance is made easier, but the risk of underperforming 'paper-OECMs' is increased. The overall benefit of loss of the implementation flexibility cannot yet be assessed.

2.3.4   Roles and expected outcomes

The roles and expected outcomes of OECMs in general (and fishery-OECMs in particular) stem, historically, from Aichi Target 11, now updated and expanded in GBF Target 3, and they apply equally to MPAs and OECMs. For OECMs, these properties are now crystallized in the Decision 14/8 definition of OECMs, their identification criteria and their common guiding principles. Notwithstanding, a recurrent question in capacity-building meetings has been about the roles and outcomes expected from OECM, in a nutshell. Decision 14/8 (Annex III, paragraph C2), based on Aichi Target 11, identified the following roles.

- To contribute to at least 10% coverage of protected area and OECMs by 2020 (previously in the 2010 Aichi Target 11). The target has been updated to at least 30% coverage in the 2022 GBF Target 3 to be achieved by 2030. Sectoral OECMs were intended to help countries reach their targets if they could identify effective OECMs These and earlier global conservation

coverage targets have often been criticized as not accurately reflecting progress in conservation because of the unequal quality in implementation of MPAs and OECMs and of their performance.

- To generate long-term positive biodiversity outcomes; to complement the role of MPAs in conservation; and to improve ecological representativeness and connectivity; including through their integration in wider conservation networks, landscapes and seascapes. This second and parallel role aims to ensure the broad effectiveness expected from OECMs (and increasingly for MPAs) in producing the expected benefits. The specific role of fishery-OECMs, and all sectoral OECMs, from that perspective is to help identify and account for the biodiversity conservation benefits or co-benefits possibly produced by their area-based management measures, even if established for more harvest-oriented purposes.[11]

- To draw attention on the important roles, rights and values of Indigenous Peoples and local communities over the biodiversity and territories they use, govern, manage and conserve, hopefully equitably. The point is extensively made in the 2018 Decision 14/8 guidance and fully reflected in the 2022 GBF Target 3. This significant role has implications for small-scale fisheries, many of which, but not all, are co-managed by the community and the State.

OECMs open additional opportunities that need to be fully appreciated, including to:

- incentivize a better general and site-level co-ordination between fisheries and conservation;
- illustrate the diversity of spatial measures taken by governments for conservation, beyond MPAs;
- provide protection for species and areas that are neither rare nor threatened;
- improve the operational application of the ecosystem approach to fisheries;
- incentivize RFMOs to increase attention on broader fisheries impacts;
- fill gaps in marine spatial planning and support ecological connectivity.

Decision 14/8 (Annex III, § C2b) also states that since OECMs are diverse in terms of purpose, design, governance, stakeholders and management, they will often also contribute to other global biodiversity targets such as those of the 2030 Agenda for Sustainable Development Goals (SDGs), or other multilateral environmental agreements.

Other roles and expected outcomes of OECMs are addressed in the following guiding principles. These principles tend to be underplayed in the literature on OECMs although they are essential to understand the criteria and the structure of the voluntary guidance. They should be considered all together, like the criteria, as they often overlap and intend to support each other.

In the sections below, the original text from Decision 14/8[12] is given as formal reference and shaded. Key words or expressions in the principles are underlined.

2.3.4.1   Principle (a): Significant biodiversity value

OECMs have a significant biodiversity value, or have objectives to achieve this, which is the basis for their consideration to achieve Target 11 of Strategic Goal C of the Strategic Plan for Biodiversity 2011–2020.

ABOUT BIODIVERSITY VALUE. The term 'biodiversity' refers to the definition of biodiversity in Article 2 of the Convention: 'the variability among living organisms from all sources including, inter alia, terrestrial, marine and other aquatic ecosystems and the ecological complexes of which they are part: this includes diversity within species, between species and of ecosystems'.

The term 'biodiversity value' is not used in the CBD criteria but is referred to in principles (a), (c) and (f). The criteria instead refer to biodiversity 'attributes' and provide a non-exhaustive list of examples of such attributes, which include communities of rare, threatened or endangered species, representative natural ecosystems, range of restricted species, key biodiversity areas, areas providing critical ecosystem

functions and services, and areas for ecological connectivity (CBD Decision 14/8 (2018), Annex III, Criterion C).

In addition, the first preambular paragraph of the Convention (United Nations, 1992b) refers to the ecological, genetic, social, economic, scientific, educational, cultural, recreational and aesthetic 'values' of biological diversity and its components, reflecting the broad range of values that could be considered in identifying and assessing OECMs. These values are linked to those of ecosystem functions and services which are discussed below but, as the IPBES regional assessments documented in depth, the 'values' of ecosystem goods and services are very different among cultures and economies (IPBES, 2018a, b, c: Annex II). The 'value' of an ecosystem's structural properties (e.g. in terms of species assemblages and habitats) depends on their role in supporting ecosystem functions. The work of IPBES is attaching high priority to developing consistent foundations for appropriately inclusive approaches to assessing biodiversity 'values' taking the diversity of natural ecosystems and human cultures into account, with the Values and Valuation Assessment (IPBES, 2022b) exploring this complex issue thoroughly and receiving consensus support from IPBES Member States. In marine systems the parallel CBD initiative on Ecologically or Biologically Significant Marine Areas (EBSAs)[13] can be very useful in identifying areas of high biodiversity value, in fishing areas, as in the case of the Disko Fan OECM in Canada, as well as the sponge OECMs and the New England, Corner Rise and Orphan Knoll seamounts OECMs in the NAFO regulatory area (see Box 2.1).

More recently, the economic and financial values of biodiversity have been considered in the framework of biodiversity credits. This approach is limited, however, not only because individual biodiversity components can be valued in different ways in different contexts and cultures (Mace *et al.*, 2012, in Wauchope *et al.*, 2024), but also because biodiversity is not just one homogeneous entity but a web of diverse, dynamically interacting components. In any case, methods and information attaching financial value to interactions are rarely available, and the interactions are sufficiently

Box 2.1. Northwest Atlantic Fisheries Organization (NAFO) deep-sea sponge and seamount OECMs

Kenchington, E.[14] and Benediktsdóttir, B.[15]

The ICES/IUCN-CEM FEG Workshop on Testing OECM Practices and Strategies (WKTOPS) (ICES, 2021), held in March 2021, was the impetus for the Northwest Atlantic Fisheries Organization Scientific Council (NAFO SC) to consider the suitability of submitting its areas closed to protect vulnerable marine ecosystems to the World Database on Protected Areas (WDPA). NAFO scientists participated in WKTOPS, which included the NAFO Sponge Closures and the NAFO Seamount (Corner Rise) Closure as two of six case studies used to evaluate areas with spatial fisheries measures in place as OECMs, aided by IUCN/CEM/FEG guidance on OECMs (Garcia *et al.*, 2021). While WKTOPS did not draw conclusions on whether their case studies should be submitted as OECMS, they did review the information available to support such a submission, including a brief description of each case study, highlighting particular aspects of the case study that emerged as being unique or important to the OECM evaluation. In July of that year, the outcomes of WKTOPS were presented to the NAFO Joint Commission-Scientific Council Working Group on the Ecosystem Approach Framework to Fisheries Management (WG-EAFFM).

At the NAFO Annual Meeting in September, the Commission decided, as recommended by WG-EAFFM, to establish an informal group of managers and scientists to: (i) evaluate current NAFO VME closures and other relevant management measures against the OECM criteria, and (ii) consider the implications of presenting NAFO's VME closures and any other relevant management measures to the CBD as possible classification as OECMs (NAFO, 2021). The work was postponed due to workload issues until 2022 when members of WG-EAFFM took on the preparatory work. In June 2023, the NAFO Scientific Council reviewed the evidence base for OECM submission for the sponge and seamount closures (NAFO, 2023a) and found that they met the OECM criteria as defined in the CBD Decision 14/8 (CBD, 2018). In August 2023, based on WG-EAFFM Recommendation 3a, the Commission requested the Secretariat, in consultation with the Scientific Council as required, to submit the seamount closure areas and the sponge VME closures 1–6 to the CBD Secretariat and to the UNEP-WCMC for inclusion in the World Database on OECMs (NAFO, 2023b). The Commission adopted the WG-EAFFM Report and recommendations in September 2023 (NAFO, 2023c) (Fig. 2.1), providing the NAFO Secretariat direction on OECM submission.

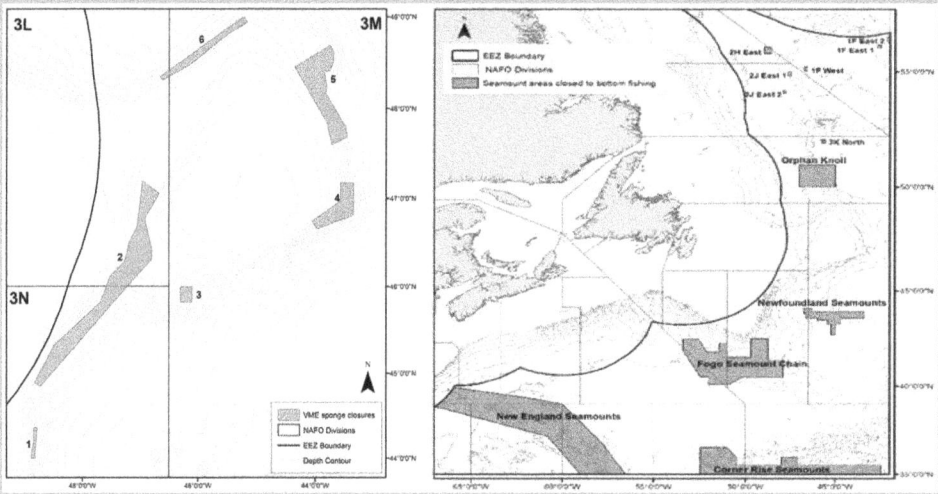

Fig. 2.1. (Left panel) NAFO VME sponge closures grouped in one OECM. (Right panel) Current areas closed by NAFO to protect seamounts (grey polygons). Image courtesy of NAFO.

Continued

Box 2.1. Continued

In NAFO, deep-sea sponge grounds are dominated by high biomass of large structure-forming demosponges that reach sizes of more than 25 cm diameter and constitute more than 99% of the total invertebrate biomass over extensive areas (Murillo *et al.*, 2012) (Fig. 2.2).

Fig. 2.2. Underwater image of the sponge grounds on Flemish Cap showing the rich diversity associated with these habitats. Photo credit: Fisheries and Oceans Canada. Used with permission.

An increased level of biodiversity has been shown to occur in these sponge grounds (Murillo *et al.*, 2020a) which provide significant functions important in delivering ecosystem heath and services, such as water quality and biogenic habitat (Pham *et al.*, 2019; Murillo *et al.*, 2020b). As a consequence of connectivity modelling (Kenchington *et al.*, 2019; Wang *et al.*, 2024) the six NAFO sponge VME closures were considered as a single interconnected system which collectively sustain significant regional populations of sponge and their associated ecosystem functions. Based on connectivity evidence, NAFO considers the sponge closed area complex as a single OECM. Further, 12 seamount areas have been closed to bottom-contact fishing: the New England, Corner Rise, Fogo and Newfoundland Seamount Chains, the Orphan Knoll and a cluster of seven individual seamounts in the northern part of the NAFO Regulatory Area, and were deemed to meet the criteria for OECMs. Different fisheries measures have been put in place by NAFO to protect these seamounts and associated species since 2006. In February 2025, the NAFO Secretariat submitted the sponge and seamount closures to UNEP-WCMC. The process is now under way to evaluate the coral and sea pen closures as OECMs.

dynamic that any such 'values' would change frequently, whether areas and biodiversity are used or protected.

There are also technical challenges associated with the capability of current technologies to measure biodiversity across scales, and in demonstrating the cause of any measured unit of biodiversity gain or loss (Wauchope *et al.*, 2024). At least until the present time, aesthetic, cultural and moral values of biodiversity have not been sufficient to 'bend the curve' of biodiversity loss. Harnessing the economic and finance sectors for the biodiversity cause may require a more economic definition of biodiversity value, crystallizing its commodification[16] to facilitate the use of economic incentives for transformational societal change, but such efforts open new sets of risks and debates (Bamford, 2024).

In practice, for OECMs used in fisheries, the biodiversity attributes of direct relevance referred to in this book as 'biodiversity attributes of concern' are those attributes other than the target species that are: (i) impacted by fishing operations, and for which conservation regulations are expected to eliminate, reduce or mitigate the impact, and eventually restore healthier conditions; or (ii) identified as a conservation priority, by a legitimate local, national or international authority. The attributes concerned may include non-target species listed as protected, endangered or threatened (PET) species, as well as critical or essential habitats, and may be part of the 'other important local values' of traditional community authorities, such as Indigenous Peoples and local communities (i.e. as emblematic or sacred species) and part of the 'other locally relevant values' mentioned in the OECM definition (and criterion D).

Many biodiversity attributes of concern can be routinely identified and effectively monitored in a fishery management system. For by-catch, accurate taxon identification may require special training and vigilance in catch monitoring, particularly for benthic macroinvertebrates and rare species of high conservation priority – including species of fish or seabirds. Viruses, bacteria, phytoplankton and micro-benthos species may also be affected by fisheries (e.g. by bottom trawling) but they are rarely brought on board by fishing and hence not monitorable with fishery-provided data. Where potential impacts on such taxa are a concern, monitoring and evaluation of impacts are likely to require additional funding for directed, fishery-independent study and modelling, and may be assessed only occasionally.

Some changes in species composition of an ocean area may be practically irreversible, such as those resulting from (i) the opening of new pathways for species to enter an area (e.g. man-made canals); (ii) anthropogenic changes to the physical structure of habitats, increasingly common as coastal development and watershed runoff of land-based sediment and pollutants alter coastal ecosystems; (iii) changing oceanographic conditions linked to climate change; and (iv) voluntary or accidental introduction by navigation (fouling, ballast water) or aquaculture. Such species may become functionally significant in the reconfigured ecosystem, as predators, prey or ecosystem engineers, and

even come to provide new ecosystem services to people, e.g. supporting new fisheries, such as king crab in the Bering Sea or multispecies fisheries in the eastern Mediterranean. As such, they may be sustainably used by fisheries and considered as part of the evolving biodiversity values in the areas concerned.

The aim of this principle (a) is to ensure that the positive effects of the OECM on biodiversity are measurable and large enough. The term 'significant' is undefined but it can be expected that the OECM contribution should be measurable (in absolute or relative terms) compared to some baseline reflecting the state of biodiversity values on the OECM area or in the fishing ground when the OECM status was granted and/or before, e.g. in case restoration is needed. A measurable impact may be small or large and the principle does not specify how big the outcome should be in order to be considered 'significant'. Moreover, a small benefit to a particularly threatened species may be much more 'significant' for reaching healthy biodiversity than a much larger benefit to a population already large and secure.

Fishery-OECMs, like ABFMs, are likely to vary in terms of the degree to which they will positively benefit biodiversity, both overall among different types of OECMs and even for the same type of OECM applied in different ecological and socioeconomic contexts. Hence if degrees of biodiversity benefit were considered, it would be prudent to assess benefits case by case, avoiding generalizations. Moreover, given the information reviewed in this section, decisions on how significant the benefit should be in order to deserve being considered a fishery-OECM would also have to made on a case-by case basis.

ABOUT PRIMARY AND SECONDARY OBJECTIVES. Fishery-OECMs usually have two broad objectives: (i) to ensure the conventional sustainability of the fishery; and (ii) to generate long-term benefits, particularly for the biodiversity features of concern. The broad qualitative objective of 'conservation of biodiversity values' may be the primary or secondary objective of OECMs (Decision 14/8, Annex III, 1c, 1d; see also Section 2.3.2 above). If an existing ABFM is identified as an OECM, optimization of fishery benefits while maintaining productive stocks size will often remain the primary objective. The current or incremental biodiversity conservation benefits could be a secondary objective, but one that would now have to be monitored and sustained if the fishery-OECM status is to be maintained, as is necessary, for example, when conventional ABFMs are designed for conservation of sensitive habitats and vulnerable species, and are intrinsically good potential OECMs (e.g. in VMEs).

These 'objectives' are quite broad and little more than expected qualitative properties, and no specific target and reference value for any biodiversity value are given in Decision 14/8. This could not be done at global level and for all the biodiversity but it implies that the effectiveness of an OECM and its management may be evaluated only qualitatively. However, the legitimate authority can establish more quantitative targets and indicators, as part of its monitoring and evaluation system (see Chapter 7).

2.3.4.2 Principle (b): Complementary conservation role

OECMs have an important role in the conservation of biodiversity and ecosystem functions and services, complementary to protected areas and contributing to the coherence and connectivity of protected area networks, as well as in mainstreaming biodiversity into other uses in land and sea, and across sectors. OECMs should, therefore, strengthen the existing protected area networks, as appropriate.

The aim of this complex principle is to stress that the main role of OECMs is to contribute to the *in situ* conservation of biodiversity, complementing the role of MPAs, ensuring coherence, improving connectivity (see also principle (d)) and mainstreaming biodiversity across sectors. This discussion accordingly builds on Section 2.3.1 in which PAs and OECMs properties are compared

COMPLEMENTARITY. Referred to in Decision 14/8 Annexes I and II, and principles (b) and (d), complementarity between MPAs and fishery-OECMs is interconnected with the other additional properties of OECMs discussed in this book: representativeness, connectivity and integration within conservation networks. This property can be viewed from two different perspectives.

On the one hand, OECMs are subsidiary, with the biodiversity benefits and conservation coverage from OECMs being incremental to those provided by PAs. On the other hand, OECMs are essential to fill major gaps left by the limitations on fully protected areas to allow humanity to receive ecosystem services essential for wellbeing. Regardless of debates between the two perspectives, the effective combinations of OECMs and PAs provide the best outcomes for both biodiversity and humanity, with 'effective combinations' yet again being case specific.

In practice, complementarity could be objectively materialized by (i) common conservation objectives; (ii) similarities or complementarity in the biodiversity values being protected; (iii) optimization of the location and size of the areas concerned (positive adjacency); (iv) harmonization of the respective technical regulations; and (v) co-ordinated gap-filling in functional conservation networks, improving connectivity. It could be argued that OECMs might also play a complementary role to MPAs by providing protection to some species/habitats not yet protected by MPAs, or where local opposition to MPAs is strong or the governance processes to establish MPAs does not exist, while sectoral conservations measure may have community support. Developing complementarity may require additional 'institutional bridges' and specific collaboration between fishery and conservation institutions and science.

The assessment of complementarity takes account of whether the OECM fills a gap in the biodiversity attributes protected in the ecological network, strengthens the functional connections among the network's areas, or manages pressures or threats in ways that allow measures in other areas in the network to be more effective. The concept implies that the parameters used to assess the biodiversity conservation outcomes of OECMs should be comparable to those used to assess MPA outcomes. In addition, the respective outcomes should complement each other (additionality) or enhance each other (synergy). This implies that wherever MPAs exist in the vicinity of OECMs, connectivity pathways are identified (e.g. life cycles, major connecting drivers, etc.).

At the very least, OECMs should not provide biodiversity outcomes that would conflict with the objectives of MPAs to which they are functionally connected. Where an MPA could only provide partial protection to key biodiversity attributes (e.g. if the species migrates outside the MPA for part of its annual life history cycle), additional protection provided by the neighbouring OECMs would be useful. Moreover, the fact that an MPA could be made more effective by adding complementary regulations in neighbouring areas could provide additional incentives for establishing OECMs in these areas.

OECMs' complementarity may be realized in providing additional protected 'stepping-stone' areas in the life cycles of protected species, or in protecting critical habitats or food sources for these species. As appropriate, complementarity with other areas defined in the ocean for biodiversity-related purposes, such as Ecologically and Biologically Significant Areas (EBSAs) or Key Biodiversity Areas (KBAs), might be considered.

Operationalizing complementarity may require additional 'institutional bridges' and specific collaboration between fishery and conservation science. Complementarity in area-based networks might be shown by the similarity in the biodiversity elements protected in OECMs and nearby MPAs (positive adjacency). It can be enhanced by connectivity, for example through migration and diffusion of life-stages, even if the areas were physically some distance apart. OECMs should complement other existing area-based conservation measures: (i) adding verified biodiversity benefits either through their own direct biodiversity outcomes or enhancement of the effectiveness of other network areas; (ii) increasing the area coverage of the network; and (iii) improving or filling gaps in representativeness and connectivity.

However, complementarity cannot be an important feature of fishery-OECMs until functional networks of MPAs have been established, within which these OECMs could show a complementary role. However, it could be argued that OECMs might also play a complementary role to MPAs by providing protection to some species/habitats not yet protected by MPAs, where local opposition to MPAs is strong, or the governance processes to establish MPAs does not exist while sectoral conservation measures may have community support.

MAINSTREAMING. The legal foundation for the concept of biodiversity mainstreaming can be found in CBD Article 6(b), under which Parties are required to 'integrate, as far as possible and as appropriate, the conservation and sustainable use of biological diversity into relevant sectoral or cross-sectoral plans, programmes and policies'. The Global Environmental Facility's Scientific and Advisory Panel (GEF-STAP) defines 'mainstreaming' as: 'The process of embedding biodiversity considerations into policies, strategies and practices of key public and private actors that impact or rely on biodiversity, so that it is conserved and sustainably and equitably used both locally and globally (Huntley and Redford, 2014).

This simultaneous mention of 'complementarity' (in reference to protected area networks) and 'mainstreaming' (in reference to sectoral management) in the same principle highlights the two-way intent of OECMs to (i) facilitate the use of sectoral tools to advance biodiversity conservation and (ii) promote the explicit consideration of biodiversity objectives in sectoral management.

2.3.4.3 Principle (c): Dual role in in situ conservation

OECMs reflect an opportunity to provide in situ conservation of biodiversity over the long-term in marine, terrestrial and freshwater ecosystems. They may allow for sustainable human activity while offering a clear benefit to biodiversity conservation. By recognizing an area, there is an incentive for sustaining existing biodiversity values and improving biodiversity conservation outcomes.

Related to principle (a), this principle reflects a key difference between MPAs – generally established primarily for protection purposes – and OECMs, particularly those used in fisheries. Fisheries-OECMs based on existing ABFMs will traditionally have the sustainability of the fishery and targeted resources as primary objectives yet may also produce long-term positive biodiversity outcomes. In contemporary fisheries, even conventional ABFMs established for sustainable use of fishery resources are increasingly expected to contribute to reducing the fisheries' ecological footprint and complementing/strengthening existing conservation networks. To meet these expectations, spatial and non-spatial tools are increasing introduced or adapted, primarily to strengthen the positive biodiversity outcomes. The fact than an area-based measure aims to serve both traditional sectoral objectives (maintaining the resource base and deciding who can fish where and when) and broader biodiversity objectives (generating long-term positive biodiversity outcomes) could raise doubts and suspicion in both fisheries and conservation arenas, when both arenas see these objectives as in tension. But that dual purpose is also a golden opportunity to give a concrete sense to the term 'sustainable use', defined by the CBD (United Nations, 1992) as: 'The use of components of biological diversity in a way and at a rate that does not lead to the long-term decline of biological diversity, thereby maintaining its potential to meet the needs and aspirations of present and future', and reflected in the contemporary framing of sustainability in IPBES (2022a).

Finally, this principle stresses that the identification does not aim simply at identifying already effective OECMs (among pre-existing ABFMs). The principle is an incentive to enhance the biodiversity performance of existing OECMs by strengthening or introducing additional technical regulations or creating new OECMs, and provides an incentive to both recognize additional ABFMs that also have conservation as a prominent priority objective (possible for FMPs in many VMEs), and include biodiversity considerations whenever spatial measures are considered for an FMP.

2.3.4.4 Principle (d): Outcomes of comparable importance

OECMs deliver biodiversity outcomes of comparable importance to and complementary with those of protected areas; this includes their contribution to representativeness, the coverage of areas important for biodiversity and associated ecosystem functions and services, connectivity and integration in wider landscapes and seascapes, as well as management effectiveness and equity requirements.

COMPARABLE IMPORTANCE. The expression 'comparable importance' allows broad interpretation, as it can take into account both the magnitude of the biodiversity outcome and its nature, and

this breadth of interpretation means individual decisions can be contested. The concept of 'importance' has multiple dimensions and two different areas may be of 'comparable importance' for biodiversity even though one area may be particularly important for, say, connectivity, while another may be particularly important for 'management effectiveness and equity requirements'. This can be a crucial consideration in comparing a potential OECM and an existing formal protected area or, indeed, comparing across OECMs.

Interpreting the 'magnitude' of outcomes faces the challenge that a change in such outcomes may be perceived as quite minor by some perspectives but important by others, depending on how they value biodiversity or specific biodiversity attributes or features. The 'nature' of an outcome can be described objectively, but similar challenges are encountered when deciding the relative degree of 'importance' of the biodiversity component (e.g. between a mud flat and rocky habitat or between a species of zooplankton and a species of whale), and the seriousness of the risk incurred if the OECM was not put in place. Moreover, 'magnitude' and 'nature' inherently interact, because even a small incremental benefit may be 'important' if it contributes to alleviate a serious conservation concern, whereas a larger benefit may be needed for society to consider it 'important' where no specific conservation concern has been identified and the biodiversity benefits are scattered among many species or habitat features. As with other principles, the necessary interpretations can benefit from tools like the EBSA criteria helping to guide case-by-case decisions (Johnson et al., 2024).

INTEGRATION. Integration, though not contained in the OECM definition, is extensively referred to in Decision 14/8, where it involves integration (i) of the OECM governance and management; (ii) within and between sectors; (iii) of OECMs and MPAs within ecological networks and seascapes; and (iv) of data and information. Integration is referred to in Decision 14/8 operational paragraphs 1 and 4; Annex I which is totally dedicated to the subject; Annex III, particularly in principle d, criterion C; and Annex IV. In this book, 'integration' is briefly addressed in Section 5.5.5c and in more detail in Chapter 6.

In principle (d), the integration issue is related also to 'connectivity' and 'complementarity' with MPAs within landscapes and seascapes. Integrated coastal area management (ICAM) and marine spatial planning (MSP) and other spatially integrative management frameworks would facilitate such integration. ICAM is specifically cross-sectoral and beyond fishery management mandate but the FAO (1996a) has published guidelines for the integration of coastal fisheries into ICAM. MSP principles and tools might be applied at sector level to integrate OECMs among them and with MPAs at EEZ level. Compliance by the fishery sector with this aspect of OECMs may require the direct intervention of the State to put in place and co-ordinate the implementation of a cross-sectoral legal and policy framework

Fishery-OECMs are ABFMs with an additional OECM status, adding management requirements for broader biodiversity conservation to those aiming at the sustainability of the fishery itself. Their management should therefore be integrated in the FMP to facilitate synergy between the regulations inside and outside the OECM (see below).

Since the elaboration of FMPs became a formalized best practice, ABFMs have logically been integrated in such plans. However, the existing or new expected biodiversity outcomes of the ABFM that the OECM status requires and their operational implications need to be referred to the management plan of the fishery for which they have been designed. For example, the OECM requirements will be reflected in the FMP in the form of additional objectives and regulations required to achieve them, as well as additional enforcement, monitoring and evaluation means, both inside and outside the OECM if required. For example, if a gear is prohibited in the OECM to protect a by-catch species like a turtle, it might be necessary to require a turtle excluder device, a by-catch ban or some economic incentives in the fishery outside the OECM to avoid the benefit produced inside the OECM being reduced or cancelled by the same fishing activity outside it.

Further, the OECM status may require a higher level of integration if several fisheries operate in the same sector and their plans need to be co-ordinated to secure the positive biodiversity outcomes. Such co-ordination must at least manage the risk that one fishery operating in an area or on a target species substantially reduces or eliminates the OECM benefits expected from another carefully managed fishery in the same area or another connected one. In addition, such integration of OECMs across multiple FMPs may create additional synergies among their OECMs, improving connectivity and effectiveness. In addition, both a sector-wide and cross-sectoral OECM perspective might be *integrated* with each other and with MPAs in the area, in cross-sectoral 'landscapes' (i.e. in inland waters) or seascapes set up by the State, optimizing *connectivity* under broader spatial planning frameworks such as ICAM or MSP.

However, in cases where a formal FMP is not yet in place (as is often the case in small-scale fisheries), a specific management plan for the OECM would need to be established scaled to the capacity available in the coastal community.

2.3.4.5 Principle (e): Demonstrated outcomes

OECMs, with relevant scientific and technical information and knowledge, have the potential to demonstrate positive biodiversity outcomes by successfully conserving in situ species, habitat and ecosystems and associated ecosystem functions and services and by preventing, reducing or eliminating existing or potential threats and increasing resilience.

DEMONSTRATED OUTCOMES. Demonstrating a biodiversity outcome is a necessity for OECMs but also a challenge as for all management tools used in complex socioecological systems. The use of the 'best scientific evidence available', to manage resources, reduce fisheries footprint and increase resilience, is required by UNCLOS and when implementing the ecosystem approach to fisheries (EAF) (FAO, 2003). In complex and dynamic socioecological systems, unequivocally demonstrating the impact of a single measure (be it area based or not) is a high-order challenge. Moreover, through the 21st century the contribution of other knowledge systems, including the knowledge of Indigenous Peoples and other local communities, to 'best available information (or evidence)' has been recognized (Diaz *et al.*, 2016) and is captured in many of the GBF Targets, while all CBD processes are working to broaden their knowledge foundations.[17] Greater collaboration between conservation and fishery science would help to address these challenges.

THREATS. 'Threats' are mentioned here for the first time but are referred to many times elsewhere in the Decision, as a constant concern, particularly under criterion C1 (see Section 2.4.3.1). Threats may be existing, current, new, potential, anticipated or pervasive. 'Preventing, reducing or eliminating existing, or potential threats' is seen as a condition to obtain and maintain the expected biodiversity and ecosystem outcomes. In order to be addressed, threats need to be identified with their nature, source or cause and impact and then managed, prevented, reduced, eliminated and addressed, preferably collectively rather than individually to address their potential interactions.

A range of current and potential impacts of various fisheries on biodiversity have already been identified in many reviews.[18] It may also be feasible to assemble information on threats from other sectors active – or planning to be active – in the area of the potential fishery-OECM. However, some case-specific information will be needed to document at least the general magnitude of the potential risk[19] that these threats represent in the area concerned and around, were the fisheries-OECM not in place. This would help prioritize and optimize the preventive or corrective regulations that may be needed. Once these regulations have been in place for an appropriate period (which depends on the biodiversity outcomes expected), it should be possible to present evidence that the risk is being reduced sufficiently for the biodiversity outcome to be secure if the regulations remain in place. When, on the contrary, the area-based measures are new, the justification for OECM status may be argued from experience elsewhere or through simulations, but should then be supported by specific monitoring or other information collection systems able to provide information over time on how the risk is being managed.

Principle (e) also states that OECM management is consistent with the ecosystem approach and the precautionary approach, and adaptive.

2.3.4.6 *Principle (f): Representativeness and connectivity*

OECMs can help deliver greater representativeness and connectivity in protected area systems and thus may help address larger and pervasive threats to the components of biodiversity and ecosystem functions and services, and enhance resilience, including with regard to climate change.

This principle complements principles (b), (d) and (e) on the relationship between OECMs and MPAs and the effect on resilience. It considers 'representativeness' and 'connectivity' as two properties that may help address large external stresses on biodiversity like climate change, but potentially also land-based pollution, coastal degradation and demography.

Representativeness and connectivity are, without any doubt, useful properties to optimize biodiversity conservation, and this remains valid even under adverse external drivers. However, two considerations can be determined: (i) larger and pervasive threats typically must be addressed at multiple scales and with mixes of diverse management regulations. The role of individual OECMs (whether or not they are in a designed network) in addressing such threats may be conceptually argued (as is done for MPAs) but difficult to formally demonstrate; (ii) the external drivers can modify the ecosystem to the point that any regulations taken at some point in time to ensure representativeness and connectivity may need to be changed later to maintain either or both of the two functions.

CBD Decision IX/20 (2008, Annexes II and III)[20] provides scientific guidance on criteria and steps regarding the design of ecologically representative and well-connected networks of MPAs, which could also be applicable to OECMs.

Representativeness and connectivity are considered in more detail below.

REPRESENTATIVENESS. Ecological representativeness is referred to in Annex II and principles (d) and (f). It refers both to the biodiversity values contained in the OECM site compared to those contained and possibly threatened in the larger ecosystem, and to the contribution of the OECM to the conservation network in that ecosystem.

Accounting for ecological representativeness has not been a major consideration for conventional ABFMs as their characteristics were specifically aimed at protecting specific target stocks. Assessing the ecological representativeness of the fishery-OCEM biodiversity in a large biogeographic unit represents a significant broadening of the concepts of 'impact' and 'management' to scales larger than that of the stock, the personal experience of fishery actors, and their local knowledge, and possibly the mandate of fishery managers and institutions.

Representativeness is a fundamental quality of MPAs' networks that indicates that the network protects representative samples of all species in the ecosystems present in the area covered by the network, at a sufficient scale to contribute effectively to their long-term persistence. Principle (f) specifies that OECMs may improve the *representativeness* of existing MPA networks e.g. in terms of presence/absence of major habitat types, key natural resources and ecologically important areas and processes.

This principle is at the core of the commitments within the CBD. The concept of representativeness has been part of environmental assessment and protection for some time, and guidance on how to apply the concept is available (e.g. Bourgeron *et al.*, 2001; Rice and Houston, 2011). Representativeness is always scale dependent, and the phrase 'larger and pervasive threats to the components of biodiversity and ecosystem functions and services' indicates this should be evaluated at least at the scale of the ecosystem concerned.

Inclusion of representativeness as an additional property was intended to encourage both that individual conservation areas (including MPAs and OECMs) were large enough to allow important ecological processes to occur naturally within them, and that spatial protection did not focus solely on rare and unique areas but ensured that typical ecosystems within larger geo-ecological regions had areas of enhanced protection for their biodiversity. On larger scales, overall representativeness can only be achieved by networks of areas, supported by effective gap analysis at network level, to identify shortfalls in

protection and locate protected areas so that the network covers much of the biodiversity attributes in need of protection (Dudley and Parish, 2006; Williams *et al.*, 2016).

Ecological representativeness of fishery-OECMs, therefore, relates to the range of biodiversity attributes of concern[21] in the ecosystem – such as threatened populations, life-stages and vulnerable habitats – that reside and are protected in the OECM. The OECM representativeness contributes to the overall representativeness of the conservation network in the ecosystem, and depends on the extent to which it fills biodiversity protection gaps in such networks (e.g. as biodiversity 'banks', corridors or stepping stones). This role is particularly important in coastal areas where no-take areas are becoming ever harder to accept. Principle (d) clarifies that the contribution of the biodiversity available in an individual OECM to a network representativeness should consider the scale at which the relevant conservation authorities are developing their networks of conserved and protected areas. Within such networks, effectiveness of individual fishery-OECMs could be improved if several fishery-OECMs used in the same ecosystem could be merged or harmonized for maximum impact.

CONNECTIVITY. Connectivity is referred to in Annexes I, II and III (subcriterion C3) and we focus here on connectivity between OECMs and MPAs in conservation networks.

In ecology, connectivity may be defined as a measure of the extent to which the movement between habitat patches of biodiversity features like eggs, larvae, juveniles, adults and populations is facilitated by the network structure and the dynamics of the surrounding environment. In area-based conservation, it is a property of the network of areas, with the habitats and species they host, which enhances the flow of energy and biomass across the network, maintaining or enhancing biodiversity (Rudnick *et al.*, 2012).

Connectivity depends on species and context and can be affected by various human activities. It may be maintained primarily by avoiding as much as possible the creation of barriers across migration or transport corridors. In a spatial conservation network, it can be enhanced by optimal geographical location of the spatial measures and spatial gap filling. Connectivity can be hampered by environmental heterogeneity and degradation including fragmentation of essential habitats and population structures. It can be a significant concern for migratory and amphidromous species (like salmon or eel) when barriers block or degrade their migration corridors. The problem may be more particularly acute on the seafloor than in open waters (based on Worboys, 2010; Rudnick *et al.*, 2012). Physical barriers to migration may be created *inter alia* by river dams, filling of wetlands, coastal development, heavy pollution and some fishing techniques, particularly in fresh and brackish water systems, although many fishing gears are usually prohibited in narrow straits. Actual physical barriers blocking marine migrations in the open ocean are uncommon although the role of intensive fishing is not well known. Addressing this concept in the ocean is a non-trivial task requiring tight collaboration between conservation and fishery science, and oceanographers and marine biologists.

Connectivity may be structural and functional and both types may have vertical and horizontal dimensions.

Structural connectivity between OECMs and MPAs depends on their shape, size and adjacency as well as the physical and oceanographic connections among them. The relevance of these properties varies with the species, their behaviour (e.g. pelagic or demersal) and life-stages (e.g. larvae, juveniles, adults). In the open ocean, the connectivity 'corridors' are strongly dynamic and three-dimensional. Areas with complex biodiversity assemblages have different structural connectivity pathways on the bottom and in the water column, and the corridors will be used differently by diverse species at different life-stages. Structural connectivity may be modified more easily by fishing on the bottom (e.g. by trawling on biogenic habitats) than in the pelagic domain.

Functional connectivity relates partly to the extent to which the structural connectivity of OECMs and MPAs facilitates the three-dimensional movements of biodiversity components in the completion of their life cycle. In addition, functional connectivity can relate to the linkages of mobile predators to patches of prey, provision of refugia, etc. For example,

protection of forage fish in an OECM might enhance the feeding of protected seabirds and their reproduction in a distant protected rookery. In the context of food webs, functional connectivity results in trophic cascades between trophic levels.

The resulting connectivity among OECMs and MPAs in a network is a complex 3-D phenomenon that must take into account ecological and oceanographic knowledge but, even with good information, is hard to quantify and monitor (Goulletquer *et al.*, 2013). There are underlying structural aspects of connectivity that can be measured by the size, composition and vicinity of OECMs and other protected areas, and major oceanographic features (currents, upwellings, etc.) that can be measured. However, the oceanographic features themselves are seasonally and interannually dynamic, and the actual functional connectivity further depends on what happens between conservation areas (e.g. fishing pressure, noise pollution, contamination, predation) and how those intervening factors affect movements and energy flows.

Horizontal and vertical connectivity. In the tridimensional ocean, connectivity may also be seen as vertical and horizontal. Horizontal connectivity can be reduced through partial obstruction (e.g. gears like pelagic drift nets at the surface on bottom-contacting species on the seabed). It can also be reduced by direct interception of migrating features by fisheries in areas that are neither OECMS nor MPAs, and through harvesting, reducing the sizes of the populations moving between areas.

The various elements of the 3D structural connectivity in the marine environment play different roles for different species, facilitating or obstructing functional connectivity, depending on the mobility of the species and life-stages concerned. Consequently, it must be considered at different scales, from neighbouring reefs (e.g. for feeding or mating) to very large oceanic gyres for completion of a whole life cycle (e.g. in the case of tunas, turtles or lobsters). Connectivity through migrations and feeding behaviour can make biodiversity components receiving protection from fisheries in one place or depth stratum vulnerable to fisheries when they move to areas or depths where the protective regulations do not apply. In addition, maintaining or improving ecological connectivity between areas under different jurisdictions has international implications and thus may require multilateral agreements for straddling or transboundary OECMs.

Although field studies on OECM connectivity have not yet been published, a simulation of VME connectivity through Lagrangian particle tracking explored the impact of removal of habitat patches in the North Atlantic, to determine the relative importance of each patch to larval connectivity within natural networks settings (Wang *et al.*, 2024).

There is significant literature on ecological connectivity and key references may be found in Meiklejohn *et al.* (2010).

2.3.5 OECMs governance and management

2.3.5.1 *Principle (g): Consultation*

Recognition of OECMs should follow appropriate consultation with relevant governance authorities, landowners and rights owners, stakeholders, and the public.

Stakeholders and 'rights-holders' or 'rights owners' are referred to in several places across Decision 14/8. In Annex II (B, 9, footnote 23), they are defined as follows

In the context of protected areas, 'rights holders' are actors with legal or customary rights to natural resources and land, in accordance with national legislation. 'Stakeholders' are actors with interest and concerns over natural resources and land.

In addition, in Annex (IV, C, 2) which is dedicated to the subject, it is suggested that: (i) relevant rights-holders and stakeholders should be identified, considering livelihoods, cultural and spiritual specificities at various scales.

The introduction of any new management measures, spatial or otherwise, increasingly requires some degree of participation of stakeholders and rights-holders in modern fisheries management set-ups, from inactive receivers of information to active participants in knowledge collection, monitoring, assessments, decision making and enforcement. The latter model of active participants (high-level participation)

is usually assumed to arise in traditional fisheries management but in fisheries broadly, decision-making authority is now increasingly being shared, through co-management and co-creation of management approaches, or even becoming fully devolved governance (Grafton *et al.*, 2009; Yates, 2014; Jentoft, 2017; Leite *et al.*, 2019; Dawson *et al.*, 2021; Puley and Charles, 2022; Govan *et al.*, 2024). The crucial need for participatory approaches is also well established with establishment and operation of MPAs, from which lessons are available for OECMs (e.g.Charles and Wilson, 2009).

In contrast to the above moves to promote active participation of stakeholders and rights-holders, the term 'consultation' used in this principle allows weak 'participation' of stakeholders, with decision making reserved for top-down authorities. However, experience in many contexts has shown that to create a responsive and responsible management system, the management strategy, plans (including rebuilding plans) and precautionary decision rules[22] ought to be discussed and, ideally, agreed upon by the main stakeholders to ensure improved buy-in and compliance. This principle of consultation includes these forms of greater empowerment of resource uses and other stakeholders and rights-holders, but does not require them in OECMs.

2.3.5.2 *Principle (h): Legitimate governance capacity*

Recognition of OECMs should be supported by measures to enhance the governance capacity of their legitimate authorities and secure their positive and sustained outcomes for biodiversity, including, inter alia, policy frameworks and regulations to prevent and respond to threats.

The *legitimate authority* is the authority with the formal mandate (traditional or given by the State) for decision making in an area or a sector. In marine fisheries, it is the State, or any authority mandated by the State such as a ministry of fisheries. Depending on the context, reference may also be made to legitimate governance or management authority. In densely developed areas with spatially overlapping uses, an OECM may require several collaborating legitimate authorities. These authorities need means to exercise their mandate and, in fisheries, tend to be chronically underfunded.[23]

Capacity building is therefore a constant and increasing need as States and other legitimate authorities face continuously growing and more complex challenges as they consider a growing number of dimensions and drivers. Specifically, for OECMs, a broader biodiversity-oriented capacity will be needed in fisheries management, the additional cost of which might be reduced by a stronger collaboration with the ministry, or other relevant authority, responsible for biodiversity or research already active or interested in the ecosystems and biodiversity (see Table 4.2).

Compliance is affected by the sense of *legitimacy* of the decision-making authority. In the ocean, the State is the only recognized legal authority, although in ABNJ States may collectively exercise their authority through RFMOs. Within their EEZs, a State may partition legal authority for different parts of the marine realm and its uses to different ministries, and particularly in coastal waters, the overall authority might be decentralized, formally devolved (e.g. to local communities, fishing associations, rights-holder groups and municipalities) or recognized as traditionally held by such communities (e.g. in the case of Indigenous People). As with several other principles, on a case-specific basis, a proposed OECM must explain and take into account how the State has established the *legitimate* authorities to deal with the various aspects of biodiversity and its threats, including fishery operations, that are relevant to the outcomes of the fishery-OECM.

2.3.5.3 Principle (i): Indigenous Peoples and local communities

Recognition of other effective area-based conservation measures in areas within the territories of Indigenous Peoples and local communities should be on the basis of self-identification and with their free, prior and informed consent, as appropriate, and consistent with national policies, regulations and circumstances.

This principle refers to situations in which OECMs may affect fisheries or biodiversity components recognized as important to the wellbeing or livelihoods of Indigenous People and local communities. The joint reference to Indigenous Peoples, on the one hand, and 'local communities,' on the other hand, is a

common one in international fora. There is a sense that, in management and conservation regulations, careful consideration is needed if both, in terms of possible impacts of those regulations on Indigenous Peoples broadly and on vulnerable local communities of all forms. That said, in situations in which Indigenous Peoples are involved, there is a particular need to ensure (i) self-identification and (ii) free, prior and informed consent, since many past initiatives took place without these being considered. In that regard, OECMs are no different from other management and conservation measures.

The idea is that if the area is to be considered an OECM, there must be 'free, prior and informed consent' of the relevant Indigenous People, cultures and communities. Increasingly often, this principle is accompanied by governance, including management responsibility, being devolved to Indigenous Peoples and other local communities or recognized by the State under traditional or modern area-based use and management rights. In such governance settings, it would be the devolved governance body that has the right to establish fishery-OECMs, if so desired, although the State may encourage and facilitate such actions. This recognition of traditional rights could facilitate OECM development and identification in community-managed areas like 'locally managed marine areas' (LMMAs) in the South Pacific and elsewhere, as well as in areas with 'territorial use rights in fisheries' (TURFs[24]), such as in Chile. Within these areas, the competent local authorities might decide whether to establish OECMs.

2.3.5.4 Principle (j): Cultural and spiritual values

Areas conserved for cultural and spiritual values, and governance and management that respect and are informed by cultural and spiritual values, often result in positive biodiversity outcomes.

This principle highlights a key aspect of OECMs, that their primary objective may not be biodiversity conservation, yet they may be very effective vehicles for such conservation. There are many ways in which this effectiveness arises. First, the sustainable use of biodiversity implicitly requires that it be carefully conserved; this highlights the strong linkage between communities, conservation and livelihoods (Charles, 2021). Second, cultural and spiritual values inherently reinforce that link of conservation and sustainable livelihoods, in that the harvesting behaviour of people reflects their values, including their bonds with nature (Charles *et al.*, 2024). Third, cultural and spiritual values may also lead to conservation quite separately from aspects of sustainable use – sacred sites and taboos, for example, may support conservation based on areas and practices that are 'off limits' in the particular culture.

The principle of accounting for values other than the ecological and socioeconomic ones is addressed in Section 2.4.4.2 in relation to criterion D2. This principle may apply mainly in inland and coastal, small-scale, fishing communities more than offshore and in the high seas. While the implicit focus may be on Indigenous Peoples around the world, the principle can apply also to non-indigenous settings where the cultural and spiritual values may be entrenched in centuries of fishing traditions, sometimes in offshore areas far away from home.

2.3.5.5 Principle (k): Governance systems

OECMs recognize, promote and make visible the roles of different governance systems and actors in biodiversity conservation. Incentives to ensure effectiveness can include a range of social and ecological benefits, including empowerment of Indigenous Peoples and local communities. [emphasis added]

The first part of this principle, while not worded clearly, suggests that OECMs must allow for 'different governance systems and actors'. This is really no different from any intervention since multiple forms of governance may already be applied in fisheries ranging from authoritative top-down management by the State to complete devolution to coastal communities, municipalities and fisheries associations. In the marine fishery sector, OECMs may be implemented under all of these forms of governance, although 'private governance', as property *sensu stricto*, is extremely limited in the ocean.

The second part of the principle is important in highlighting the idea of 'incentives', related to the benefits accruing from the OECM (which also relates to equity considerations of who receives those benefits). Participants in

fisheries are likely to consider themselves to be bearing many of the costs of adopting and maintaining the fishery-OECMs. This is similar to the situation with conventional ABFMs but in the latter case, the benefits to fishers and the fishing community may be clear in terms of improved target resources, more sustainable livelihoods and better fisheries profitability. In the case of biodiversity benefits in fishery-OECMs, however, the additional costs of the biodiversity benefit are most likely to be paid largely or fully by the local fishing community, whereas the resulting benefits will be distributed to a larger community, across the region, the country and globally, potentially altering the sense of equity. Inequities need to be fixed to achieve both buy-in and compliance by the fishers and their associated communities, and for achieving the expected biodiversity outcomes.

For these reasons, incentives are often important as part of developing fishery-OECMs and networks. These can be monetary incentives, such as expectations of higher revenues if the target resources also benefit from the additional measures, or eco-certification of fisheries with effective OECMs. They can also be non-monetary incentives such as more effective local empowerment in governance and increasing local ownership of the measures. This can facilitate local innovation to reduce economic costs and increase awareness of the expected social and cultural benefits for the community, and produce a better image of the fishery in the media and popular dialogue.

2.3.5.6 Principle (l): Best available information

The best available scientific information, and indigenous and local knowledge, should be used in line with international obligations and frameworks, such as the United Nations Declaration on the Rights of Indigenous Peoples, and instruments, decisions and guidelines of the Convention on Biological Diversity, for recognizing OECMs, delimiting their location and size, informing management approaches, and measuring performance.

The need to identify, collect and use local knowledge and best-informant stakeholders, together with scientific knowledge, is progressively gaining more traction. This principle is particularly important in relation to indigenous or local ecological and technical knowledge (Berkes, 1999; Diaz *et al.*, 2016). Use of such knowledge is not only needed when formal scientific knowledge is unavailable or very scarce, but even when there is plenty of scientific material, since indigenous and local knowledge may be superior or may at least fill gaps. In fisheries, a huge amount of information is contributed by fishers on catch (often including by--catch and capture of protected species), fishing effort and their spatial distribution, and much of that information is used in contemporary assessments. The degree to which their broader ecological knowledge is collected and used is patchy.

Underlying these considerations about the need to use all forms of knowledge is the matter of equity, specifically of equitable access to, and ability to use, various forms of information. That is made clear in the Kunming-Montreal Global Biodiversity Frame (GBF) which has as its fourth main goal:

> Adequate means of implementation, including financial resources, capacity-building, technical and scientific cooperation, and access to and transfer of technology to fully implement the Kunming-Montreal Global Biodiversity Framework are secured and equitably accessible to all Parties...

This is reiterated in GBF Target 21 (on 'data, information and knowledge for decision making') which calls for ensuring that 'the best available data, information and knowledge are accessible to decision makers, practitioners and the public'. As highlighted by the IUCN-CEM Fisheries Expert Group (Rice *et al.*, 2023):

> The Target places emphasis *inter alia* on the best information available, its accessibility, its role in equitable governance, the participation of stakeholders, the management mandates, and the use of Indigenous Peoples and local communities' knowledge.

In line with the United Nations Declaration on Rights of Indigenous Peoples[25] (UNDRIP) (United Nations, 2007), free, prior and informed consent (FPIC) centres on obtaining consent from Indigenous Peoples (IPs) for any activities undertaken on their land. At the basic level, governments cannot implement a policy or programme on or concerning Indigenous Peoples'

lands unless there is prior consultation and consent from the Indigenous community[26] (see Section 5.5.2.2). It is not yet clear whether, in international law, FPIC is also applicable to 'local communities' and when collecting information in the fisheries sector and small-scale fisheries communities, but it is promoted for wide application in the KM-GBF and associated CBD guidance. Once this local or indigenous knowledge is mobilized, methods for integrating the often more narrative knowledge of IPs and other local communities and fishers with the quantitative approaches of modern fisheries assessments are still in exploratory stages and need further testing and development (Butler *et al.*, 2012; Tengö *et al.*, 2017; Cowie *Et al.*, 2020; Klara loch and Riechers, 2021; Chambers *et al.*, 2021; Lopes de Susa *et al.*, 2022; Charles, 2023).

Aside from the provision of information, the degree to which fishers are also directly and formally involved in the assessment and elaboration of management advice (directly or through their representatives) is also very important for the buy-in and compliance required for proper planning and implementation of conservation and sustainable use.

2.3.5.7   Principle (m): Transparency and evaluation

> It is important that OECMs be documented in a transparent manner to provide for a relevant evaluation of the effectiveness, functionality and relevance in the context of Target 11.

Transparency is required not only for data sources, assumptions used in assessments, confidence limits of such assessments, and identification and evaluation of management options, but also for the processes used for objective setting, decision making and evaluating management performance, as well as for the rationales and justifications for the outcomes of those processes. Without such full transparency, the credibility of both the decisions regarding fishery-OECMs and the evaluation of their performance in relation to all their objectives can be questioned. Among the important implications of this principle is the need for an appropriate monitoring system, a recurrent assessment process (the periodicity of which depends on the biodiversity attribute being monitored), full and timely release of the monitoring information and assessment findings, and some system of oversight (or audit) ensuring the quality and reliability of the monitoring, assessments and adaptive decision making, increasing the level of trust among the complex set of stakeholders.

2.3.6 Relation between OECM and EAF principles

Decision 14/8 refers to the ecosystem approach as a key framework in its Annexes I, III (principle (e) and criterion B4c), and IV. It is clear that application of the guiding principles to fishery-OECMs will be greatly facilitated in fisheries already managed within an ecosystem approach, revising it as needed to better match the OECM specifications. CBD decisions V/6, in 2000, and VII/11, in 2004, contain detailed guidance on the implementation of the ecosystem approach, and which therefore should be observed in the context of OECMs.

More specifically to the fisheries sector, and complementary to the CBD guidance on the ecosystem approach, Annex 2 of the FAO Guidelines on EAF (FAO, 2003: 83–88) lists a more detailed set of relevant principles: (1) avoiding overfishing; (2) ensuring reversibility and rebuilding; (3) minimizing fisheries impact; (4) considering species interactions; (5) ensuring compatibility of regulations between jurisdictions (for shared or straddling resources); (6) applying the precautionary approach; (7) improving human wellbeing and equity; (8) allocating user rights; (9) promoting sectoral integration; (10) broadening stakeholder participation; and (11) maintaining ecosystem integrity. As one would have expected, OECM guiding principles above are in line with the above principles 2, 3, 4, 5, 6, 9, 10 and 11 with more specification regarding the differences and complementarity between OECMs and MPAs.

There is also a close parallel between the 'human dimensions' needed for OECMs and those inherent in the EAF (see, e.g. de Young *et al.*, 2008). The latter includes such considerations as (i) the relevant policy, legal and institutional frameworks, (ii) institutional and legal frameworks affecting EAF, such as the good governance principles, and (iii) social

and economic considerations, which include employment, livelihood and regional aspects, distributional aspects (winners and losers), poverty and social security/services, food security, cultural and religious considerations, and Indigenous People and traditional knowledge. Among the approaches common to both are (i) the assessment of benefits and costs, and (ii) the use of suitable environmental and social impact assessment and decision-making tools. Both OECMs and EAF, and indeed MPAs, also need to consider incentive mechanisms that can support or oppose their implementation.

2.4 Criteria for Identification and Evaluation

The identification of existing or new area-based fishery management measures (hereafter, ABFMs) that might deserve the OECM status is based on the four criteria for identification and ten subcriteria provided in Decision 14/8 (Annex III), and available in Appendix 2 of this book. The criteria (numbered A–D) and subcriteria (e.g. C1–C4, etc.) reflect many of the 'properties' of the area-based conservation measures already noted in the 2010 Aichi Target 11 and later in the 2022 GBF Target 3. For each of them, the Decision indicates also some 'elements of evidence' (e.g. C1a–C1d) to consider for a positive assessment (Appendix 2, column 2). This numbering convention will be used throughout this book. The same set of criteria will be used to assess the OECM performance during its long-term implementation. Consequently, the term 'area' used when reviewing the criteria below may refer to an existing or new ABFM at the inception of the OECM identification process, or to an operational OECM during the performance evaluation process required to maintain the OECM status later on.

When going through the criteria, it will be noted that they have been listed practically in the same order as the related properties in the definition. Although it is often stressed that all criteria are equivalent and should be given the same level of attention, it should be stressed that OECMs are, first of all, geographically delimited 'areas' with particular properties. In addition, some of the criteria are inherently binary (present or absent) and thus mandatory pass or fail. Others are more qualitative, with the degree to which they are met or not being judged by the evaluators.

Criterion A, related to the legal status of the area considered as a potential OECM, is binary and eliminatory. If it is not met, the identification fails and the area cannot be considered as a potential OECM even if it meets most other criteria.

Next, subcriterion B1 asks for a geographical definition, and is a condition to report OECMs to the national authorities and the CBD, to register an OECM in the World OECM database managed by WCMC and to calculate its coverage. Subcriteria B2 and B3 are respectively about the governance and management action in the OECM. For areas fulfilling criterion A, subcriteria B1, B2 and B3 provide the basis for qualitative evaluations of governance and management processes. Minimum standards to be met are specified in that governance is *legitimate and appropriate* for *achieving* the objectives. The same qualities of governance are required from the management system. Both have to involve relevant authorities and stakeholders, sustain *in situ* conservation of biodiversity, be consistent with the ecosystem approach, be adaptive and have the ability to manage threats. However, the level of sophistication in governance and management in subcriteria B2 and B3 is unspecified. This allows scope for later improvement in governance and management, without negating immediate eligibility for OECM status if all other criteria are met. This also allows areas with strong community-based decision making and management, using customary practices rather than institutionally proscribed procedures, to be considered for OECM status.

Subcriteria C1–C3 are about the current or expected outcomes of potential fishery-OECMs and are also aspirational. As long as their intent is not clearly violated in the potential OECM, those subcriteria may not be completely met at the initial assessment, and as long as better outcomes can be reasonably expected, progress towards the outcomes will be a consideration in the periodic performance reviews.

The management regulations and the wording of the elements of evidence B3a and C1a overlap completely, ensuring consistency in application. Subcriterion C4 stresses the importance of monitoring and evaluation as well as information archiving. These aspects were not explicit in Target 11 and the OECM definition but extensively referred to in the other elements of the CBD guidance.[27] They are qualitative (how much monitoring and evaluation is adequate may be case specific) and, importantly, are properties that must be maintained by consistent practices of the jurisdictional authority, and not by the ecosystem features. Subcriteria D1 and D2 are about relevant values in the OECM other than biodiversity, such as ecosystem services and functions, and other locally relevant social and economic values (of importance to biodiversity conservation). They too are qualitative, in the sense that no specific services or functions, nor local values, are specified as essential. They may better be viewed as criteria that can exclude areas that meet the other criteria but where ecosystem services and functions, or local values, are being actively degraded. However, just providing ecosystem services and functions, or local values, does not make an area an OECM if other criteria are not met.

In principle, although not specified in the Decision, all criteria are to be 'considered' in identification and performance assessment, meaning that all criteria mut be evaluated.[28] This notion is reinforced by the principle of transparency regarding the documentation and evaluation of OECMs as discussed above (CBD Decision 14/8, Annex III, A and Principle (m)). A criterion is considered to have been 'met' when the relevant available information has been duly considered and that as a minimum the area does not violate its intent. However, some criteria might be irrelevant for a particular area, e.g. traditional cultural values for areas located on a seamount of the Mid-Atlantic Ridge. The absence of traditional cultural values would not make the area non-eligible. The criteria would nonetheless be *considered* and a rational explanation would therefore available as to why it was considered irrelevant in the case concerned.

On the contrary, when drafted, the 'elements of evidence' were not intended to be indicative and not mandatory or exhaustive, and may vary significantly depending on the specific site and context. They are part of the voluntary guidance provided in the Decision, and might be flexibly interpreted, increased or enriched, nested or combined, keeping in mind the need to remain consistent with the intent of the Decision.

Although failing to comply with a criterion may be a good reason to reject the potential OECM being examined, failing to satisfy an element of evidence among the many others that are relevant for a subcriterion can only affect the overall score that such subcriteria will reach (see Section 5.6.1). These standards can be complex to apply when an ABFM offers clear benefits to some but not all elements of evidence for a criterion. When only some of the elements of evidence are relevant for a specific ABFM, and they are satisfied, the scoring is straightforward. However, when some key elements of evidence benefit substantially from the ABFM, whereas other elements actually suffer harm from the fishery, the scoring involves balancing the need for the positive outcomes with the ability of the negatively impacted populations or ecosystem properties to sustain the associated impacts.

These criteria and subcriteria are reviewed in more detail in Chapter 5 about the identification process. In the following sections, short elaborations are offered for each criterion and subcriterion.

2.4.1   Criterion A: The area is not currently recognized as a protected area

The area is not currently recognized or reported as a protected area or part of a protected area; it may have been established for another function.

CBD Aichi Target 11 and GBF Target 3 allow each jurisdictional authority to report all areas identified as legally protected, or culturally protected for Indigenous Peoples as contributing to their jurisdictional targets, but specifically exclude 'double counting' any area in the global coverage. Hence, even if a protected area was found to meet all OECM criteria, it could not be reported if it was already reported as a protected area. Hence the CBD Decisions and processes offer no incentive for determining whether the ABFM being considered meets MPA criteria – or a legally protected area meets any or all of the OECM criteria. From the perspective of the CBD area coverage Targets, any time an area has been already designated as an MPA or part of an MPA (e.g. a buffer area), and reported as such by the legitimate authority to the CBD or the WCMC World Database on Protected Areas (WCMC-WDPA), there is no need for further action. However, none of the CBD guidance *prevents* evaluating candidate OECMs against MPA criteria, or MPAs against OECM criteria, if the authority itself has internal reasons for doing so and does not report the area to the CBD twice under different statuses. In fact, according to the WCMC User Manual for the World Database on OECMs (UNEP-WCMC, 2019), potential OECM areas may encompass areas that meet the definition of a protected area, in cases where the governance authority *prefers* the area to be considered an OECM.

The simplest way to address criterion A would be to clarify the legal status of the area and ensure that it has not been designated as an MPA by the legitimate authority (see Section 2.4.2.2). One should also check whether the area is listed as an MPA in the WCMC-WDPA, and hence potentially used already in the global coverage accounting. However, the WPDA contains numerous areas which are not counted in the global coverage (e.g. biosphere reserves). In addition, some areas (like Ramsar wetlands) may not be considered MPAs by some States and not reported in their national PA statistics, generating discrepancies between national and WCMC statistics. Conversely, some States report on the WPDA some Ramsar sites that do not meet the MPA management criteria. In some countries, however, designations of some types of areas (e.g. cultural areas) require explicitly no overlap with MPAs, formally resolving the issue.

Importantly, the criterion stresses that the area-based measure considered as a potential OECM may have been established initially for another function, with different primary objectives, stressing from the onset that the conservation objective of an OECM does not have to be conservation, consistent with section 2.2 of this chapter. There is also the possibility that a formally registered but partially failing MPA might be delisted and identified as an OECM as decided by a State if it meets the criteria or can be improved to meet them.

When dealing with existing ABFMs, established under the authority of a fishery authority, conversion of some or all of such an area to legal status as an MPA usually involves mandatory consultation and agreements across ministries, so the risk of overlooking or being unaware of such a change in status is very low. However, the risk may be higher if fisheries jurisdictional authorities and other ocean managers rarely communicate, or if national policies allow MPAs to be established without consultation with industry sectors and government agencies that could be affected by the MPA.

2.4.2   Criterion B: The area is governed and managed

It is important to stress at this point that OECM status is granted to an area (identified by criterion A) and not to any system of governance or management or any collection of technical regulations. The need to maintain such area in the long term is only implicit in Decision 14/8 but obvious and fundamental as discussed further below.

The overall aim of this criterion is to avoid 'paper OECMs' that would not produce the expected outcomes in the absence of good governance and effective management. Criterion B has three subcriteria: (i) the area is geographically defined; (ii) it has legitimate governance authorities; and (iii) it is managed. These will be briefly examined below.

2.4.2.1   Subcriterion B1: Geographically defined space

(a) Size and area are described, including in three dimensions where necessary. (b) Boundaries are geographically delineated.

One consequence of subcriterion B1 is that the fishery-OECM can be localized on a map, possibly with co-ordinates, and its area should be stated or could be calculated to be accounted for in Target 11. A clear geographical location is a condition for the OECM to be accepted in the world OECM database (UNEP-WCMC, 2019:16). The WCMC user manual of the OECM database[29] recognizes (i) preferably a set of geographical co-ordinates of the boundary, e.g. in a GIS shapefile; (ii) a single or multiple polygon; or (iii) the latitude and longitude of the centre-most

point of the area. The depth over or below the delimited area is apparently not requested when registering an OECM, although this subcriterion B1b mentions explicitly the *third dimension*. Similarly, the eventual subdivisions of the OECM into horizontal subareas or vertical strata is not mentioned in Decision 14/8 but highly relevant (see Section 2.4.2.3c).

In a broad sense, *all* fishing activities are spatially structured by preferred fishing grounds related to the presence of preferred target resources, within which fishing vessels freely move and operate. These fishing grounds are usually not legally designated areas and hence do not meet criterion B1. In theory, fishing grounds could be mapped as the historical footprint of the fishery, delimiting historically used areas, and could be endorsed by the State (as in South Africa) to constrain future expansion of the fishery. Currently, however, conventional fishing grounds cannot be considered as ABFM areas and hence as potential OECM areas.

Similarly, ocean space is often divided into fisheries management units for statistical record keeping (as FAO statistical divisions) or to delineate jurisdictional competence (e.g. of RFMOs), and such management units also are not, in themselves, appropriate for consideration as potential OECMs.

Incidentally, it can be noted that a string of small neighbouring and functionally connected ABFMs might be considered as one single OECM (as for VMEs in NAFO; see Box 2.1), consisting of a mosaic of non-overlapping subareas, for management and reporting convenience. This approach would also combine the benefits from all subareas, cashing in on the benefits generated by their connectivity, for example if each subarea could not meet all criteria but the mosaic of areas could.

STATIC   VERSUS   MOBILE   BOUNDARIES. Marine resources tend to move seasonally across the tridimensional ocean space, some of which on very long distances, across specific benthic or pelagic areas, for the completion of their life cycle, e.g. for reproduction, early growth, recruitment, maturation, reproduction and survival (e.g. refuges). Benthic or pelagic ABFMs are often put in place to protect these places. These ABFMs may be fixed when they are always the same (e.g. when related to bottom habitats or structures), or mobile when related to environmental conditions and poorly predictable. In all these ABFM areas, the fishery regulations may be permanent, seasonal,[30] temporary or changed in real time (Rice *et al.*, 2018). Stable areas with permanent fishery regulations may also be established to allocate resources and sometimes management responsibilities to specific coastal communities or to reduce inter-fisheries conflicts.

In all these ABFMs, the fisheries regulations may generate some broader collateral (unintended) biodiversity benefits, and additional conservation regulations may be put in place to generate or enhance such outcomes. Decision 14/8 does not explicitly deal with the issue but provided the biodiversity outcomes are demonstrated (criterion C), the principles and criteria lead to the following considerations.

i.   *Temporary ABFMs*, in which both the area and all the regulations are short-lived (related to a stock rebuilding period), do not meet the sustained management requirement (Subcriterion C2) and could not be OECMs. If the fishery regulations end with the rebuilding management regime but the conservation regulations are maintained, the area would become a fixed OECM with conservation as primary or sole objective (see case (ii)).

ii. *ABFMs with fixed boundaries* and permanent regulations are likely to be the best potential OECMs. If regulations were only seasonal, ABFMs could be considered potential OECMs if there is sufficient evidence of long-term benefits despite the only seasonal protection.

iii. *Real-time ABFMs* for which both the area boundaries and the regulations move on short time scales (from days to months) would not have a stable defined geographical boundary and would fail on subcriterion B1. For seasonal migrations, the whole migration area might be considered as one fixed ABFM (case (ii) above) but this would raise considerable issues regarding: (a) whether to consider as ABFM the whole migration area or only the area with highest concentration on average; (b) in the last case, what threshold abundance would

be used to draw the boundary? Would the biodiversity benefit and OECM performance be the same or oscillate seasonally?

iv. *Drifting ABFMs* would change position slowly under climate change, probably in the same direction, to maintain resources protection. This issue is examined below.

The necessary mobility of ABFMs (and their OECM status) as a consequence of progressive change in climate and the related change in resources and biodiversity distribution and parameters is serious. A decrease in average fish abundance of 3.4–24.1% is expected by the end of the century, with increases in higher latitudes and decreases in tropical regions. It is expected that, by 2030, almost 25% of transboundary stocks will have shifted their distribution across maritime jurisdictions, and distribution shifts are projected to continue at a rate of tens to hundreds of kilometres per decade, reducing the effectiveness of spatial management measures. Similar phenomena are affecting the broader ocean biodiversity even though less continuous information may be available. Heatwaves, for example, have already impacted ecosystems and living habitats (like kelp forests and coral reefs), fisheries and fishing communities (Zuo *et al.*, 2015; OECD, 2024).

In theory, ABFMs may be relocated from time to time to slowly adapt to the climate-driven changes in the fishery resources conditions and distribution. The degree of biodiversity protection offered by the area between changes in location may decrease with time but the timing of the stable periods may be adjusted to specific local conditions. During the climate-driven shifts, species leaving a place are likely to be replaced by others coming in. Consequently, depending on local situations, the old OECM status of the area and related regulations may be cancelled, or maintained with updated objectives. Conservation areas must consider the fact that they will need to adapt to climate change, particularly as biodiversity continues to move across jurisdictional boundaries. The World Commission on Protected Areas (IUCN-WCPA, 2019: see Box 4.1) already considers that, in exceptional circumstances, MPA boundaries may be defined by physical features that move over time, such as riverbanks, the high-water mark or the extent of sea ice.

In fisheries management, the recommended response to climate change is mainly a reduction, redistribution and decarbonization of fishing pressure. However, a stronger application of the ecosystem approach, spatial planning and area-based measures like fishing reserves or refuges and MPAs is also considered to protect vulnerable species during distribution shifts and to protect or recover essential habitats (see Barange *et al.*, 2018). In that respect, the role of MPAs in mitigating climate change has often been heralded (Roberts *et al.*, 2017) but not without controversies (Bruno *et al.*, 2018; Bates *et al.*, 2019; Hilborn *et al.*, 2021a; Smith *et al.*, 2023). In addition, the importance of combining area-based measures (including ABFMs and MPAs) with non-spatial regulations that control fishing pressure in and around them is becoming clear (Hilborn *et al.*, 2004, 2021b; Roberts *et al.*, 2017; White *et al.*, 2024).

Decision 14/8 does not foresee the reporting of continuously moving OECMs (or MPAs) and further formal guidance from the CBD COP would be helpful. Notwithstanding the contemporary importance of the issue and consideration of the available principles and criteria, the following preliminary considerations could be made.

- The location, boundary and regulations of ABFMs will need stepwise changes to maintain the protection of moving resources and biodiversity.
- The bottom-related features and living habitats will move less than the pelagic ones and may suffer more if they cannot adapt to the new conditions.
- The older locations may retain a functional role despite climate change, but for the replacing species and fisheries, with adapted fishery and conservation regulations.
- The initial biodiversity of concern may not move in the same direction as the fishery resources and at the same speed. To stay in cooler environments, some elements may move north, others south. The expected benefits might therefore need to be reassessed.
- If the shift results in an overlap between two OECMs (possibly established by two different sectors), or between an OECM and an MPA, or in case of change in a pre-existing

overlap, the 'no double counting' rule applies and the coverage may need to be corrected.

- Similarly, other economic activities and related ABMTs may not move in the same way or at all, leading to potential overlaps in jurisdictions and areas (e.g. an oil field or a deep-sea mining concession), potentially generating new threats for biodiversity and new constraints for the economic activities concerned.
- As long as the drift in resources and biodiversity remains relatively slow and small, the OECM may be maintained with occasional adjustments of the boundary. The regulations may be adapted to the shifting conditions. The nature of the biodiversity benefits, and the conservation-related regulations, may change but as long as significant benefits are produced, the OECM status may be maintained.
- If the drift becomes important, the new area may need to be redesigned and its OECM status entirely reassessed. The previous protection area might be cancelled or maintained as a new ABFM for the new fisheries and associated biodiversity.
- The resources and/or the biodiversity may split into multiple foraging or breeding concentrations, such as different gyres, during the move, and merge again later. Would the different areas be part of the same OECM? How would management costs be influenced?

These uncertainties can trigger a certain reluctance in management institutions to consider mobile MPAs and OECMs but climate changes are a certainty affecting their performance and adjustments will need to be considered, with strong scientific, operational and international implications. The eventual revisions of the OECM status are considered in section 7.8.

2.4.2.2 Subcriterion B2: Legitimate governance authorities

The evidence to be considered is: (a) Governance has legitimate authority and is appropriate for achieving in situ conservation of biodiversity within the area: (b) Governance by Indigenous Peoples and local communities is self-identified in accordance with national legislation; (c) Governance reflects the equity considerations adopted in the Convention; and (d) Governance may be by a single authority and/or organization or through collaboration among relevant authorities and provides the ability to address threats collectively.

The purpose of criterion B is to ensure there is accountable and uncontested governance and management of the area in ways consistent with CBD Decisions[31] which should reduce the probability of identifying ineffective OECMs. The requirement of formal governance and management also implies logically (and is argued by IUCN-WCPA, 2019) that if an area is not governed and managed, even in a natural or near-natural state, then it cannot be identified as an OECM until it is formally delimited, identified and regulated by appropriate authorities.

Fisheries, and the OECM identified in them, are expected to be formally under the responsibility of a mandated fishery authority (e.g. central, local, hybrid or traditional). That authority should be in charge of implementing the fishery-wide management plan and of making and enforcing the regulations needed to maintain effective OECMs, ensuring the other considerations of relevance for OECMs, such as equity in the distribution of costs and benefits of the OECM, and addressing actual or potential threats to these outcomes from fishing or other sources (see sections 2.4.2.3, 5.2.7 and 5.5.2.6 on threats and risk assessment). Because of the biodiversity conservation role of the fishery-OECM, a collaboration of the legitimate authorities of fisheries and biodiversity conservation may often be valuable, potentially increasing recognition, effectiveness and means available, e.g. for monitoring and assessment.

The responsibility for reporting at international level is not completely clear in the Decision, in which States are encouraged to report on OECMs to UNEP-WCMC and to get registered in the global World Database on OECM (WD-OECM) This parallel database on MPAs has been the historical record of progress towards the quantitative element of area-based conservation targets adopted under the CBD. However, this does not replace the formal and official national reporting to the CBD. However, CBD Parties are looking for more standardized metrics to evaluate progress towards targets,

leading to a strong focus on indicators and a systematic approach to assessing progress at the global level for the GBF implementation. This may possibly lead to a future CBD comprehensive online reporting tool. At the moment, however, the only required reporting is to the CBD through the usual national reporting process. The reporting to the WD-OECM is optional but this database will probably continue to play an important role on global reporting (J. Appiott, CBD Secretariat, personal communication).

Decision 14/8 stresses the importance of the wide range of governance systems under which OECMs may be identified and used, from centralized or decentralized State-driven governance to shared governance (e.g. co-management) and community-based governance as in the case of Indigenous Peoples and other local communities. In Decision 14/8, the term 'legitimate authority' is used (notably in criterion B1) together with the 'governance authority' and 'management authority', underlining the importance of legitimacy across all levels of governance. In marine fisheries, the only legitimate authority recognized by current international law is the State or an authority mandated or recognized by the State or established by States, at the international level (e.g. RFMOs) or the national level (e.g. national agencies, Indigenous Peoples and other local communities, municipalities and fishery associations. However, tensions exist in some countries between the centralized State's authority and customary rights-holders (see Govan *et al.*, 2008, 2024; Dominguez and Luoma, 2020), and solutions have usually been negotiated at the national level which provide for varying degrees of recognition of these rights.

It can be noted that the spaces of relevance to both sustainable use and conservation may straddle jurisdictions. Using the 1995 UNFSA terminology, a fishery-OECM might be located: (i) between two national jurisdictions ('shared' OECM); (ii) between the national and international jurisdictions ('straddling' OECMs); and (iii) between two international regional jurisdictions, adjacent or not (e.g. a regional fisheries management organization (RFMO)[32] and a regional seas convention (RSC)[33]). In relation to the preceding section on 'geographically defined space' (subcriterion B1), it is clear that no jurisdiction can define fishery-OECM boundaries outside areas under its mandate, providing an incentive for co-operation across jurisdictional boundaries.

2.4.2.3   Subcriterion B3: The area is managed

OECMs are expected to be: (a) managed in ways that achieve positive and sustained outcomes for the conservation of biological diversity. (b) Relevant authorities and stakeholders are identified and involved in management. (c) A management system is in place that contributes to sustaining the in situ conservation of biodiversity. (d) Management is consistent with the ecosystem approach with the ability to adapt to achieve expected biodiversity conservation outcomes, including long-term outcomes, and including the ability to manage a new threat.

This subcriterion is particularly important and connected with the two preceding ones. The aim with respect to fishery-OECMs is to ensure that they are effectively used in an active fishery management system. In larger-scale commercial fisheries, these management systems would include objectives, plans, enforced regulations and monitoring, and decisions would be taken and enforced in a participative and adaptive manner, both consistent with the EAF. OECMs in small-scale and community-based fisheries may or may not have similar management systems but may well draw on local culture and established sustainable practices (see FAO, 2015), often involving objectives, plans, compliance measures, monitoring and (as appropriate) participatory processes of decision making. Such local systems need recognition by the more central fishery authorities, to ensure the practices of Indigenous Peoples and other local communities are respected by the management authorities.

The connected meanings of the terms 'sustained' and 'long-term' are addressed under subcriterion C2.

The *ability to manage threats* is discussed in sections 5.6.3 and 5.7.2 on threats and risk assessment. However, it should be noted that subcriterion B3, which is focused on management, does not mention the need to identify the current and future threats to be managed (controlled, reduced, eliminated).

The need to address threats is also referred to in the elements of evidence B2d and B3d but the need to undertake a threat assessment is strangely peripheral in the central management criterion. To some extent, threats are more thoroughly addressed in the elements of evidence C1b and C1c.

Areas important for biodiversity, rich in biodiversity values but not used by fisheries and not covered by a formal FMP or set of fishery management measures, cannot be claimed as OECMs. They could, however, become an OECM, if some other societal sector (other than fisheries, e.g. military, tourism) is managing that area and producing biodiversity benefits. Further, *management authorities can make deliberate decisions to leave an area untouched* (e.g. as a reserve) (IUCN-WCPA, 2019:5), and as long as this decision is recorded and enacted, it can be considered to be 'managed' appropriately, as long as the expected 'long-term' benefits are assured.

This subcriterion does not say anything about the specific regulations, such as access rules, gear restrictions, economic incentives and disincentives, that might be taken by the legitimate authorities inside the fishery-OECM to produce the biodiversity conservation outcomes expected from the OECM status. However, informed selection and effective implementation of the appropriate suites of regulations are fundamental to avoid 'paper OECMs'. As with all CBD decisions, Decision 14/8 does not specify choices of regulations necessary to deliver the desired outcomes, to accommodate the global diversity of approaches to and capacities for implementing various types of regulations, and the diversity of factors that may affect the appropriateness and effectiveness of individual measures. Any of the spatial measures already used in fisheries to reduce the fisheries' impacts on the ecosystem, including in MPAs, and possibly new ones as technologies continue to emerge and improve, might in principle be used inside the fishery-OECM boundaries, combined and enhanced as needed to produce the necessary biodiversity benefits.

ADAPTIVE MANAGEMENT.   Fisheries management with '*the ability to adapt*' was advocated by scientists more than half a century ago (Walters and Hilborn, 1976; Ludwig and Hilborn,

1983) and modern science-based management regimes have adopted management targets that are revised every year based on fisheries outcomes. The approach gained new momentum with the adoption of the precautionary approach, following UNCED, in the early 1990s. Many fisheries management systems have implemented this as part of their adaptive approach and *de facto* both are an integral part of the ecosystem approach. In these systems, the approaches are applied to the fishery target species as well as to some non-target species for which special protection or recovery plans have been developed. In the last two decades, the closure of VMEs can also be seen as a precautionary measure.

Adaptive management may not yet be implemented everywhere in the developing world or in modern small-scale fisheries, because of capacity and operational constraints, but in some traditional management systems, adaptive management has been practised for centuries, with fishing communities deciding on their harvesting intensity and fishing practices in keeping with natural conditions around them.

Adaptive management should have the ability to anticipate, detect and respond to new emerging threats to biodiversity (see below).

ZONING. Zoning is the horizontal or vertical subdivision, through legislation or management, of an area into subareas or strata in which uses are specifically regulated, for different objectives, accounting for different conditions, to optimize the overall management effectiveness, reduce conflict or fine-tune protection in different parts of the area. The term does not refer to horizontal or vertical overlapping of different OECMs with different or identical boundaries, resulting in a potential double counting in conservation coverage targets.

Aichi Targets 11 and GBF Target 3 were negotiated without taking account the ocean depth. This third spatial dimension, particularly in deep ocean areas, is proving to be a challenge for both EBSAs (CBD, 2023b, c) and OECMs, and the CBD will need to delve further into these challenges as work to implement the BBNJ Treaty accelerates. The vertical dimension is more important in the ocean than on land because the largest part of the ocean biomass is in the water masses, not on the bottom, and different

biodiversity features and fisheries (with different impacts) may be found in different depth strata. Therefore, zoning has always been integral part of fisheries management, in MPAs (including often their atmosphere), and other conservation areas like community-managed areas. It can therefore be assumed that the zoning of management regulations is maintained when the areas are granted OECM status as long as they do not conflict with the conservation regulations and outcomes.

Tridimensional area-based management (including zoning) is essential for effective fishery management and marine conservation (FAO, 2003; Norse *et al.*, 2005; Young *et al.*, 2007). In this regard, the spatial relationship of fisheries and other sectors using the ocean would be more explicit (particularly in relation to marine spatial planning) if their historical footprints could be formally established. In the marine realm, the delimitation of different zones for the purpose of management·may, *a priori*, be horizontal or vertical depending on context. In this regard, Decision 14/8 stresses the importance of the three-dimensional nature of marine and coastal ecosystems and of OECMs (in criterion B1). The issues below regarding horizontal and vertical zoning of OECMs are not addressed explicitly in the Decision but are briefly discussed in IUCN-WCPA (2019, Box 2) and Garcia *et al.* (2021).

ABFMs are most often horizontally delimited areas, defined on the bottom or at the surface of the ocean, including implicitly an undefined part of the water column, below or above, in which special regulations apply. Decision 14/8 clearly refers to the third (vertical) dimension, but nothing in it refers to horizontal or vertical zoning of OECMs management. It was never the intent or expectation, however, that in a marine OECM the area-based measures would have to apply to the entire water column and on the substrate, if the fishery, target species and biodiversity benefits were restricted to only some strata or the ocean or on the seafloor. However, managing activities in the water column above or below the OECM, and indeed in adjacent areas, to minimize negative impacts on its expected biodiversity benefits would make sense.

We argue that if a zoned ABFM, as a whole, produces the expected long-term biodiversity benefits and meets all other CBD criteria, its OECM status should not be questioned, as long as the activities and regulations in all zones are consistent with the OECM conservation objectives. For example, in theory, turtles may be protected at the surface by banning longlines, and seagrass beds may be protected on the bottom by banning trawling, generating numerous biodiversity benefits for the whole area. However, if trawling was replaced by bottom netting, as a less damaging gear for seagrass, turtle entanglements in those nets would not be consistent with expected benefit, questioning the appropriateness of the OECM status.

Multiple situations could be imagined and the issue of vertical zoning in fishery-OECMs cannot be generalized. Concerns have been expressed in relation to the potential lack of coherence of regulations across the water column or difficulties of three-dimensional monitoring and enforcement (IUCN-WCPA, 2019, Box 2; Garcia *et al.*, 2019: 31). Pending better clarity about the intent of the CBD with regard to vertical zoning, the issue is one where the pragmatism and flexibility provided in Decision 14/8 will be needed, balancing ecological objectives and operational realities as well as fears and opportunities, case by case.

It can be noted that institutional and management zoning also exist *de facto* over the extended continental shelf where benthic resources are under national jurisdiction and those in most of the water column are under international jurisdiction. Similarly, in the high sea, the International Seabed Authority has a mandate to regulate mining of mineral resources of the seafloor, with its likely environmental impacts, but not the activities in the water column above it.

ABOUT THREATS.   A central concern across both criteria B and C is the risk assessment and management capacity needed to detect and address the forces presently impacting biodiversity, or likely to impact it in the foreseeable future. In its principles, criteria and additional voluntary guidance, Decision 14/8 frequently refers to 'threats' to the components of biodiversity, without defining 'threats'. These 'threats' are qualified as existing, current, new, potential, anticipated and/or pervasive. They need to be managed, prevented, reduced, eliminated,

addressed collectively... using policy and regulations. 'Threats' could therefore be understood as the current (or future) 'forces' that are (or might be) exerted on biodiversity components. However, the Decision refers also to current threats to biodiversity and potential threats from new and emerging pressures (Annex IV, C,1, a) which would indicate that 'threats' arise from 'pressures', that are or may not be controlled sufficiently to maintain a low likelihood of serious or irreversible harm. This interpretation can be aligned with the CBD Glossary, under 'Drivers of biodiversity loss', lists the following 'threats to biodiversity': demography, urban development, overexploitation, pollution, climate change and invasive species, apparently equating 'drivers' and 'threats'.

In order to maintain coherence between this document and Decision 14/8, we will therefore refer essentially to 'threats', current and future, and use 'pressure' only when speaking generically about activities, e.g. when referring to 'fishing pressure'. The difference between current and future threats is important for management as the ways to deal with them and the degree of urgency are necessarily different.

The clear intent of Decision 14/8 and of this step is to (i) identify the threats, either current or reasonably anticipated, that affect or will affect biodiversity in the exploited ecosystem and (ii) provide evidence that the potential OECMs (with the technical regulations taken inside and around them) have a reasonably documented capacity to reduce or eliminate the related risks that both pressures and threats represent for biodiversity conservation. The magnitude of the impacts and the resulting underlined risks depend on the nature and intensity of the anthropogenic or natural forces involved, and on how they are or will be managed.

Decision 14/8 refers specifically to 'risk' only twice (in Annex I, I, 4 and Annex I, II, B, a), in both cases referring to 'risk reduction'. The risk attached to a given threat is often estimated by multiplying the cost of its expected damage by the probability that it will materialize. Such estimates can be fairly straightforward when dealing with economic risks but are much more difficult when the risks relate to biodiversity providing non-monetary social, cultural, spiritual, identity and relational benefits to people (Pascual *et al.*, 2017; IPBES, 2022b).

INVENTORY OF CURRENT AND POTENTIAL THREATS. A first inventory of current and potential threats on biodiversity attributes of concern will aim to identify their nature, source and scale. Current threats may come from the fishery in which the potential OECM is currently operating or from other fisheries in or around it, e.g. through overexploitation, by-catch or destructive fishing practices on vulnerable habitats. They may also originate in other economic sectors of activity, marine or land based, or from climatic oscillations and change. Current threats (and potentially also future threats) add mortality, reduce biomass and possibly modify species composition and ecosystem structure, productivity, reproduction potential and resilience to environmental oscillations and change. If they emerge from the fishery sector, threats may be directly controlled with management plans and regulations. If not, they will require collaborative management across sectors. Threats may emerge from the future evolution of the specific fishery, the fishery sector, other economic sectors or other overarching drivers, and require a precautionary approach to management (e.g. through regulations that may reduce the probability or risk attached to a threat), and the elaboration of contingency plans (to implement when the threat materializes). Climate change, like many drivers, already generates current threats on biodiversity, with noticeable impacts such as displacement of stocks and coral bleaching. It also generates potential additional 'threats' as it increases with time, increasing future risks if not mitigated or reversed in time. The same applies to some extent to the demography driver.

It would be useful to rank the threats in order of assessed or assumed importance and to collect information about their trends. For benthic habitats and species, the most obvious threat is bottom trawling and regulations may be needed to reduce bottom impact (changing gears) or eliminating this technique from the OECM. This is often the measure adopted in VMEs. Complementary regulations might be needed for control and surveillance, and for monitoring of biodiversity trends.

ABOUT THE ECOSYSTEM APPROACH. From a biodiversity conservation angle, a lot of guidance is already available on the ecosystem approach to conservation from the CBD (2004) and IUCN (Shepherd, 2004, 2008). The CBD ecosystem approach is based on 12 principles and has been defined as follows.

> A strategy for the integrated management of land, water and living resources that promotes conservation and sustainable use in an equitable way. Thus, the application of the ecosystem approach will help to reach a balance of the three objectives of the Convention: conservation; sustainable use; and the fair and equitable sharing of the benefits arising out of the utilization of genetic resources. Furthermore, the ecosystem approach is based on the application of appropriate scientific methodologies focused on levels of biological organization, which encompass the essential structure, processes, functions and interactions among organisms and their environment. It also recognizes that humans, with their cultural diversity, are an integral component of many ecosystems. (CBD, 2004:6)

The IUCN guidance follows the 12 CBD principles, offers illustrations of the EA principles application through case studies, and promotes adaptive management. It also stresses that:

> ... it is never enough to consider only protected areas (PAs) when planning conservation. Other adjacent areas need to be taken into account, and not just the buffer zone. The sustainable interaction of people and biodiversity can only be developed in a larger ecosystem area, and the ecosystem approach encourages both a larger vision on the ground and an exploration of interconnections.

From a fisheries angle, the EAF (Garcia *et al.*, 2003) has been formalized in FAO technical guidelines (FAO, 2003), progressively implemented, and evaluated (Garcia and Cochrane, 2005; Sanchirico *et al.*, 2008; Garcia, 2009; Link, 2010). While obvious progress has been made in many areas (reducing by-catch of protected and other vulnerable species, and impact on vulnerable ecosystems and essential habitats) and adapting the criteria for eco-certification (MSC reference), progress is very slow in many developing countries and in some RFMOs. There is also slow progress, especially in so-called 'developed' countries, in incorporating human dimensions (e.g.social, cultural, economic, institutional) into the practice of the EAF (see De Young *et al.*, 2008; FAO, 2009c).

Fisheries that apply an ecosystem approach generally do so explicitly, with some acknowledgement that they indeed are applying it. However, if none of the major management documents for the fishery mentions the ecosystem approach, the EAF is so open-ended that a review of almost any fishery can find some aspects of the EAF that are present in management, and some either absent or in need of strengthening.

2.4.3   Criterion C: The area achieves sustained and effective contribution to *in situ* conservation of biodiversity

This criterion is fundamental as it is the one really identifying the elements that indicate whether an existing ABFM, or a planned one, produces, or is likely to produce, sufficient and lasting *in situ* biodiversity outcomes. It defines effectiveness in terms of (i) achieving expected outcomes, dealing with threats and integrating management (subcriterion C1); (ii) sustaining these outcomes over the long term (subcriterion C2); (iii) identifying important biodiversity features protected to ensure that at least some are benefiting from the measures (subcriterion C3); and (iv) documenting the area's properties, important biodiversity attributes, governance and management processes, outcomes, performance and communication strategy (subcriterion C4).

2.4.3.1   *Subcriterion C1: The area is effective*

(a) The area achieves, or is expected to achieve, positive and sustained outcomes for the in situ conservation of biodiversity (evidence C1a). (b) Threats, existing or reasonably anticipated ones, are addressed effectively by preventing, significantly reducing or eliminating them, and by restoring degraded ecosystems (C1b). (c) Mechanisms, such as policy frameworks and regulations, are in place to recognize and respond to new threats (C1c). (d) To the extent relevant and possible, management inside and outside the OECM is integrated (C1d).

Effectiveness is expressed here in terms of delivering biodiversity conservation outcomes, managing threats, maintaining management capacity and integrating management. *Effectiveness* will be achieved if specified biodiversity outcomes are produced and *sustained* in the long term, but this can be hard to demonstrate fully and document. Because no management measure can protect and improve *all* biodiversity, and direct causality is often hard to demonstrate in the ocean, the expected conservation outcomes of a fishery-OECM must be well-specified biodiversity features of concern, known to be impacted by fisheries, and whose status can be assessed with known accuracy and precision.

The subcriterion calls for evidence of *sustained outcomes*. The concern is logical: short-lived biodiversity benefits cannot be sufficient to justify an OECM status. However, particularly in the ocean, it is impossible to ensure 'constant' outcomes in a fluid, variable and changing environment. The implication is that while a flow of benefits might be maintained through adaptive management, the specific nature of these benefits, their magnitude, the specific biodiversity elements which benefit, etc. may not be maintained constant. A further aspect of achieving sustained outcomes is the crucial need for social, cultural and institutional mechanisms that can produce a suitable environment in which to 'sustain' outcomes; for example, social cohesion and local empowerment can be key ingredients leading to compliance and support for biodiversity conservation. This is considered as well in terms of effectiveness below.

Because of the dynamic natures of both oceanic and socioeconomic conditions, including climate change, some level of risk assessment and risk management is needed to anticipate environmental risks and possible major changes to the fishery, develop contingency plans to avoid or reduce impacts and, where necessary, restore biodiversity. Tools developed under the Biodiversity Impact Mitigation (BIM) hierarchy and the precautionary approach, and the decision rules in adaptive fisheries management, all have the same general objective of effectiveness, making well-established ABFMs good potential OECMs.

Effectiveness also requires dedicated institutions and policy frameworks (repeating elements of subcriterion B3). Two expressions can be stressed: '*expected to achieve*' and '*reasonably anticipated*'. The first indicates that, in a fishery-OECM, effectiveness can consider not only the benefits already occurring in the ABFM but also those expected from regulations adapted or newly introduced for improving its outcomes. At the same time, because of the uncertainties in even data-rich stock assessments, the tasks of documenting that: (i) specific positive biodiversity *outcomes* are occurring or likely to occur for new or adapted regulations; (ii) such outcomes are not being degraded by other non-fishery sources of impact or the net benefit is *positive*; and (iii) the spatial measures are instrumental in delivering the positive outcomes, are all challenging.

Applying this criterion requires sound and sufficient evidence, be it from ongoing monitoring and assessment or, when appropriate to the fishery, the knowledge of Indigenous Peoples and other local communities. Similarly, what is a 'reasonably anticipated' impending threat is itself a risk-based choice, and the more strongly local evidence or comparable precedents elsewhere suggest a risk is imminent or likely to intensify, the more effort must be invested in developing contingency plans and/or enacting precautionary regulations. In all cases, scientific support and local knowledge will be essential. A key element in effectiveness is the inherent 'buy-in' for conservation measures, with suitable social, cultural and economic incentives.

The integration of fisheries management inside and outside the OECM is fundamental and should not be a problem for fisheries with FMPs in place, because OECMs and their specific technical regulations should be naturally integrated in such plans, with all other spatial and non-spatial management measures. However, as multiple uses of the ocean continue to increase and spread more widely, the integration of FMPs with the management plans of other sectors will be increasingly necessary, to keep all the 'existing or reasonably anticipated threats' managed coherently to deliver the 'positive and sustained outcomes for the *in situ* conservation of biodiversity'.

2.4.3.2 Subcriterion C2: Sustained governance and management

(a) The OECMs are in place for the long term or are likely to be. (b) 'Sustained' pertains to the continuity of governance and management and 'long term' pertains to the biodiversity outcome.

Subcriterion C2 addresses two issues related to effectiveness: (i) the need to sustain the management effort in the future, and hence to ensure a good probability that the governance directing and supporting it will stand the test of time; and (ii) the need to ensure that the biodiversity benefits generated in the OECM will not be eroded in time. Subcriteria C1 and C2 complement but extend each other in this respect. Whereas C1 specifies that the positive biodiversity outcomes must be sustained *in situ*, C2 adds that the governance and management measures providing the positive outcomes also must be sustained, and the positive biodiversity outcomes are not just sustained from past to present but also expected to be sustained from the present into the long-term future.

It is important to stress again at this point that the OECM status is granted to an area and not to any collection of regulations applied in that area, although these are essential components of an OECM as per its definition and criteria under the CBD Decision 14/8. Therefore, while criterion C2 stresses the need to have sustained governance and management to achieve long-term outcomes, the need to maintain the defined area in the long term for the same purpose is not only implicit but fundamental. By contrast, the types and lifespan of regulations taken inside OECM areas may change as appropriate as long as the expected outcomes are produced.

Sustaining OECMs requires dedicated institutions and frameworks (see subcriterion B3). Contributions to more effective institutions and frameworks can arise from (i) mainstreaming biodiversity considerations into fisheries management, e.g. adopting a broader concept of fisheries sustainability including human and ecosystem wellbeing; implementing the Code of Conduct for Responsible Fisheries (CCRF) and the EAF; and adapting the Biodiversity Impact Mitigation to the fishery sector; (ii) establishing robust co-operation mechanisms between fisheries and conservation authorities, and with other sectoral authorities in areas where multiple uses occur or are anticipated; (iii) integrating OECMs in FMPs, and considering inside and outside outcomes together. Even in progressive fisheries where these contributing factors are present, if threats to the biodiversity outcomes are identified or reasonably anticipated from other fisheries or other industrial sectors active in the area, the necessary integration with regulations used in those fisheries or sector could pose additional challenges. The failure to integrate across such pressures could compromise the permanence of biodiversity outcomes, and lead to the area not being considered an OECM or losing its OECM status.

A key aspect of the sustainability of governance and management measures is the support for these among stakeholders. In fisheries, the entire concept of co-management (sharing the fishery management functions between government and stakeholders) arose as a means to increase the 'buy-in' of fishers for management measures. It has become clear that those involved in creating fishery management plans support those plans, and indeed oppose illegal activity threatening the plans. The same logic holds for OECMs. Accordingly, a criterion for 'likely to be… sustained' should be the acceptance by stakeholders (or rights-holders as appropriate) of the decision making and the governance system and their effective participation in that system.

The 'sustained' concept may be examined from two angles: (i) the ordinary management regime; and (ii) the rebuilding/recovery regime.

'SUSTAINED' AND THE ORDINARY MANAGEMENT REGIME. Although no systematic review of the common lifespan of an ABFM has been found, experience indicates that such lifespan can be a few days or weeks for short-term real-time closures, to a few months for a seasonal closure, up to decades or more in more permanent cases – with longer durations becoming increasingly common for protection of essential habitats and vulnerable ecosystems since UNGA Resolution 61/105. While VME closures in deep-sea fisheries adopted by RFMOs tend not to be permanent, they are expected to be renewed as necessary, based on scientific assessments and advice. A priori, short-term ABFMs (e.g. real-time and AD hoc closures) do not seem to have the OECM prerequisites. Seasonal ABFMs may be promising potential OECMs. Playing an important role as seasonal windows in the life cycle of key target species, they may generate significant long-term biodiversity benefits if they were maintained (i.e. renewed) in the long term and not wiped out

by fishing activities once the seasonal closure is over (IUCN-WCPA, 2019).

Many year-round ABFMs are intended to protect a critical life-stage of a target fishery species (such as recruits or spawners), reduce by-catch of threatened species or reduce overall fishing pressure (with variable effectiveness, particularly in the latter case). They tend to remain in place for as long as they contribute to keeping the fishery sustainable.[34] If appropriately matched to the life cycle of the species concerned and their essential habitats (e.g. for feeding, reproduction and refuge), their spatial characteristics tend to be stable. Such ABFMs would, in principle, be terminated only if a major change happened in the fishery itself, or in the life cycle of the species , reducing or eliminating their usefulness. This could happen, for example, if the threatening gear or practice were banned permanently (eliminating the risk for the species, without loss for biodiversity).

There are increasing cases when climate changes have significantly modified the distribution area of a species and its critical life-stages, reducing or cancelling the ABFM protection and potentially modifying its OECMs functions. In such cases the boundaries of the ABFM would probably need to be moved too, just to serve the primary fishery sustainability objectives, which would result in revisiting its OECM status. However, as many ABFMs initially introduced solely to contribute to fishery objectives are being used to also contribute to biodiversity objectives, these revisions are often proving complex and may be controversial.

'SUSTAINED' AND THE REBUILDING REGIME. Fishery regulations are flexible and adaptable by design, in order to maintain a dynamic regulatory system and rapidly adapt in case of poor performance or changing conditions. In a resource 'rebuilding' management regime intended to correct overfishing, total fishing pressure will be reduced and an ABFM may be established to protect recruits, e.g. banning trawling in the area (see examples in Garcia and Ye, 2018; Kenchington et al., 2018). This ABFM may also protect vulnerable habitats (like seagrass beds), generating biodiversity co-benefits, that might support the granting of an OECM status.

However, by design, a rebuilding regime is temporary, with enforced starting and closing dates.

Concern has been expressed in conservation arenas that such temporary ABFM areas, with their specific regulations, may be in place over too short a time period to provide the meaningful and sustained biodiversity conservation benefits that would justify granting OECM status (McClanahan *et al.*, 2007; IUCN-WCPA, 2019:6). The concern is logical and the risk of 'sliding back' into suboptimal states after rebuilding is also a concern for rebuilding of fisheries target stocks, and efforts are usually made to avoid that risk (see, for example, Bloor *et al.*, 2021).

OECM processes have not been in place long enough for relevant experience to have accumulated. Nonetheless, the following considerations reduce or mitigate the risk of this happening.

- The 'sustained' and 'long-term' requirements for OECMs apply to the area, its governance, its management and its biodiversity conservation outcomes. Therefore, temporary areas should be avoided and any ABFM, in any ordinary or rebuilding management regime, needs to have evidence or a formal statement that the area is established for the long term.
- If the ABFM was already operating for decades before the rebuilding regime, producing positive biodiversity co-benefits, it would be logical to reinforce it during the rebuilding regime and to maintain it in its strengthened form. The concern should not be higher than for any other ABFM.
- If the ABFM was established for the rebuilding process, and assuming there was enough information to confirm the positive biodiversity outcomes beyond the target stock rebuilding (criterion C), a formal statement of intent that the ABFM area and regulations would be maintained after rebuilding would be required before granting OECM status.
- The ABFM function of the area, supporting the fishery stocks, might be closed but the OECM function might remain, maintaining the area with its conservation objectives, particularly because it is unlikely that all species in the concerned biodiversity assemblage will recover at the same pace

and the most vulnerable ones may require longer rebuilding times than the fishery resources.

Again, many scenarios might be imagined for the termination of the resource rebuilding regime, and the elimination of a functional fishery-OECM is only one of them and possibly not the least likely in an ecosystem-based fishery management. Indeed, the rebuilding regime may be seen as an opportunity to advance the ecosystem approach, formally looking for an OECM status for the ABFMs from the onset of the rebuilding regime, integrating the rebuilding of the biodiversity with that of the stock.

2.4.3.3 Subcriterion C3: Important in situ conservation of biological diversity

Recognition of OECMs is expected to include the identification of the range of biodiversity attributes for which the site is considered important (e.g. communities of rare, threatened or endangered species, representative natural ecosystems, range-restricted species, key biodiversity areas, areas providing critical ecosystem functions and services, areas for ecological connectivity).

This subcriterion aims to ensure that OECMs contain and protect a broad range of biodiversity attributes or values in addition to fishery resources, which are obviously part of biodiversity. In general, the more strongly management for sustainable fishing or protection of biodiversity favours some biodiversity features,[35] the more likely other biodiversity features[36] will show declines. Choosing components that are 'considered important' is left to the State or other legitimate authority. However, the illustrative list included in the criterion shows that species formally recognized as threatened or endangered (e.g. in the IUCN Red List) and/or emblematic (e.g. seabirds, turtles, marine mammals), and areas like Key Biodiversity Areas and EBSAs, are 'important'. In all cases, the biodiversity attributes on which the OECM status is based need to be identified and listed, possibly with their contribution to ecosystem functions and services (see Section 2.4.4.1). Their connections with the biodiversity located outside the OECM will help illustrate ecological representativeness and connectivity and in assessing the overall net benefit, accounting for biodiversity gains and losses in and out of the OECM.

In this book, the 'important' biodiversity features or components impacted specifically by fisheries are referred to as 'biodiversity features of concern'. These are the features for which targeted fisheries management and conservation regulations taken in the OECM are intended to generate benefits. Other biodiversity features in the area of the OECM may decrease, some even intentionally (for example, if a predator control programme were necessary to allow another species to recover). But unless the declines for other biodiversity features were serious, those features were important for the integrity of the ecosystem structure and function, and the fishery regulations being used were a major cause of the decline, they would not be a major consideration in OECM status.

'*In situ* conservation' is defined in Article 2 of the CBD. For capture fisheries, the relevant part of the definition says that '*in situ*' conservation is the *conservation of ecosystems and natural habitats and the maintenance and recovery of viable populations of species in their natural surroundings*. In capture fisheries, the conservation (maintenance and recovery) of wild target and non-target species, as well as the protection of essential fish habitats (e.g. mangroves, seagrass and algal beds), is always *in situ*. In some cases, such as for anadromous species like salmon or sturgeon, their conservation may involve a combination of *in situ* and *ex situ* management of the related culture-based marine fisheries (ocean ranching) in which land-based artificial nurseries are used to enhance stocks' productivity in the wild. Stock enhancement is more widespread in coastal lagoons, lakes, and flood plains (FAO, 1989; NACA-FAO, 2000).

2.4.3.4   Subcriterion C4: Information and monitoring

(a) Identification of an OECM should, to the extent possible, document the known biodiversity attributes, as well as, where relevant, cultural and/or spiritual values, of the area and the governance and management in place as a baseline for assessing effectiveness. (b) A monitoring system informs management on the effectiveness

of measures with respect to biodiversity, including the health of ecosystems. (c) Processes should be in place to evaluate the effectiveness of governance and management, including with respect to equity. (d) General data of the area such as boundaries, aim and governance are available information.

OECMs should be well described, including all their attributes. When describing the biodiversity attributes inside the OECM, the 'value' of those attributes on multiple-value systems should be explicitly acknowledged, e.g. for ecosystem structure and function; economic benefits locally and on larger scales; and cultural and spiritual identity of nearby communities (see section 2.4.4). All these values should be considered when setting both baselines and targets for conservation outcomes, with: (i) those targets informing the development of a corresponding set of reference values (when the information is sufficient to develop them); (ii) specified trends that management systems can be accountable for delivering; and (iii) Description of what a 'healthy ecosystem' for the general area would be.

A particular note should be made concerning some requirements for information and monitoring in terms of human aspects. There may be a tendency to focus attention on monitoring (and collecting information on) direct 'biodiversity attributes', but there is a reason why this criterion refers to such topics as 'cultural and/or spiritual values', to 'the effectiveness of governance and management, including with respect to equity'. While the conservation goal is certainly about biodiversity, the achievement of that goal requires strong attention to the human dimensions – which also implies both practical actions and monitoring activity.

When an OECM is considered in an area where biodiversity has been degraded by past unsustainable practices (which may or may not have been the fisheries of current concern), baselines can be set in the present degraded state and improvements considered 'positive biodiversity outcomes'. However, targets to be reached with 'sustained' and 'long-term' positive outcomes should reflect a 'healthy ecosystem'. Each stage in recovering degraded systems and sustaining healthy ones can use these general benchmarks for evaluating success of the fishery-OECM at protecting the biodiversity values at risk from the fishery.

These elements would also provide guidance for the monitoring and evaluation system (see Chapter 7) that can be used to measure the OECM contribution to biodiversity conservation, particularly through reporting on progress towards the outcomes (i), (ii) and (iii) mentioned above. All the outcomes require dealing with uncertainties in data and monitoring results, incomplete knowledge and understanding of dynamic processes, and inherent stochasticity in natural and socioeconomic systems, but these are challenges familiar to fisheries. However, 'ecosystem health' will be additionally challenging because although the concept is widely used, there is no set of indicators or reference values agreed upon even among the broad community of experts, let alone by binding international agreements. The concept of 'health' tends to relate to maintenance of ecosystem structure and function, avoidance of significant adverse impacts (SAIs) and conservation of biodiversity within safe ecosystem levels (SELs). The latter concept is used in Aichi Target 6 in relation to vulnerable ecosystems and threatened species, but international standards are still missing (Garcia *et al.*, 2019). The SAIs concept emerged in the UNGA resolutions on VMEs (UNGA res. 61/105 in 2006 and subsequent ones), and criteria for its assessment with regard to impacts from bottom fishing are provided under the FAO International Guidelines for the Management of Deep-sea Fisheries in the High Seas.[37]

Decision 14/8 does not set standards for the burden and level of proof needed to demonstrate sufficient information on an OECM, and monitoring of its performance (see Section 5.2.9 on this issue). It calls for evidence on achievement of *positive and long-term outcomes* (C1a) and of *restoration of degraded ecosystems* (C1b), but it does not specify management and conservation targets,[38] reference values and indicators that would objectively indicate 'achievement' of the outcomes. This is in contrast to, for example, Aichi Target 6 which clearly requires that (i) fisheries have no SAIs on threatened species and vulnerable ecosystems and (ii) the impacts of fisheries on stocks, species and ecosystems are within SELs. The 2020–2030 GBF and its Target 3 are supposed to be more ambitious than the 2010–2020 Strategic Plan and its Target 11, and at least those two considerations

on SAIs and SELs should apply when assessing effectiveness of OECMs. GBF Target 5 on fisheries commits CBD Parties to minimize impacts on non-target species and ecosystems applying the ecosystem approach. While SAIs have been defined in the FAO (2009) guidelines on VMEs, SELs remain to be consensually defined (Rice *et al.*, 2018).

Moreover, despite the central importance of 'effectiveness' in OECM criteria, this property is not required when registering an OECM in the WCMC database (WCMC, 2019). Perhaps best practice on how to provide detailed evidence in the large range of conditions in which OECMs operate will emerge in the future, from Parties' implementation, varying substantially according to local conditions.

Moreover, despite the central importance of 'effectiveness' in OECM criteria, no evidence about the degree of effectiveness is required when registering an OECM (or indeed an MPA) in the WCMC databases (UNEP-WCMC, 2019), although it is always possible for the relevant authority to upload such information in the system. Currently, there is no experience on how to provide detailed evidence of effectiveness on the large range of conditions in which OECMs operate. As experience is gained from management authorities' implementation, the nature of such 'evidence' may well vary substantially according to local conditions. The authors are aware of ongoing discussions in various fora on how best to develop a standard approach to assessing effectiveness for OECMs. With the involvement of various fisheries bodies and relevant institutions, a specific guidance on fishery-OECM effectiveness assessment could be elaborated, based on the long management experience available in fisheries.

2.4.4 Criterion D: Associated ecosystem functions and services and cultural, spiritual, socio-economic and other locally relevant values

This criterion expands the range of elements to consider when assessing the effectiveness of candidate or implemented OECMs, with considerations on human dimensions of fishery-OECMs and the interaction between human and natural wellbeing from two perspectives: (i) ecosystem functions and services; and (ii) other human values. The need for this has been highlighted at various stages in the discussion above, notably stressing the need for an active and equitable participation of stakeholders in all steps of the OECM process. Moreover, the involvement of and co-operation with social scientists, legal analysts and other experts of human dimensions would also be important.

The phrasing of the subcriterion and elements indicates that while these factors (notably 'cultural, spiritual, socioeconomic, and other locally relevant values') are important to the effective performance of the OECM, the OECM itself does not have to ensure their delivery. For example, if governance overall for a jurisdiction respects the values referred to in subcriterion D2, then the OECM operating under such governance does not need to have special additional provisions. Moreover, in individual cases, the features may not even be relevant to the evaluation (see Section 2.4.4.2)

2.4.4.1 Subcriterion D1: Ecosystem functions and service

(a) Ecosystem functions and services are supported, including those of importance to Indigenous Peoples and local communities, for OECMs concerning their territories, taking into account interactions and trade-offs among ecosystem functions and services, with a view to ensuring positive biodiversity outcomes and equity. (b) Management to enhance one particular ecosystem function and service does not impact negatively on the site's overall biological diversity.

Ecosystem functions (EFs) have been defined in many sources. For example, Costanza *et al.* (1997) considered them to be *the habitat, biological or systems properties or processes of ecosystems,* i.e. the internal functioning of ecosystems and interactions between their abiotic and biotic components that transfer energy and matter between the components of an ecosystem (as in the trophic chain) and between ecosystems (e.g. through highly migratory species). Functions regulate the composition and structure as well as diversity, productivity and resilience of ecosystems and their dynamic interconnections. When an economic (or social) value can be attached to some ecosystem function, or to the result of such

function, this value is referred to an 'ecosystem service'. Ocean productivity and its distribution across the ocean trophic chains are essential functions that humans use to produce seafood, nutritious elements and the resulting cascade of revenues, jobs, leisure, household support and cultural values that in turn define human communities.

Ecosystem services (ESs) have been defined in many different but convergent ways, sometimes introducing confusion between EFs and ESs. A well-recognized classic definition is: 'the conditions and processes through which natural ecosystems, and the species that make them up, sustain and fulfil human life' (Daily, 1997). In other words, ecosystem services are 'the set of ecosystem functions that are useful to humans' (Kremen, 2005). The authoritative Millennium Ecosystem Assessment (MEA) defined ESs as 'benefits people obtain from ecosystems'. It should be highlighted that ecosystem services not only include 'direct' services such as provisioning services (e.g. wildlife harvests) and cultural services (e.g. ritual or aesthetic benefits) but also serve human wellbeing in a less obvious manner, such as regulating and supporting services, e.g. helping to maintain a suitable climate for humans (MEA, 2005). It is virtually impossible to list all the provisioning, regulating, supporting and cultural ecosystem services or the natural products (goods) that people directly consume (see MEA, 2005; Sekercioglu, 2010; IPBES, 2016).

Indeed, this wide range of ecosystem services is crucial to consider in relation to the multidimensional nature of human wellbeing. Although ecosystem services initially were viewed in the context of monetary value, the IPBES and others have developed these concepts in much more inclusive contexts. In these IPBES developments, criticism of the ecosystem services concept as being unidirectional, with ecosystems only of value because of the services they provide to the humans who use them, has led to the term largely being replaced by the expression 'Nature's Contributions to People' (NCP) to acknowledge the diversity of value systems that can be used to measure ecosystem services (Diaz *et al.*, 2016; IPBES, 2019) (see also section 5.5.4)

The concept of nature-based solutions (NbS) that emerged recently refers to 'actions to protect, sustainably manage, and restore natural and modified ecosystems that address societal challenges effectively and adaptively, simultaneously providing human well-being and biodiversity benefits'.[39] NbS 'seek to maximize the ability of nature to provide ecosystem services that help address a human challenge, such as climate change adaptation, disaster-risk reduction or ... food production'[40] (see also Cohen-Sacham et al., 2016). The concept is not without controversy and was strongly debated in the GBF elaboration process, but both the NCP and NbS concepts are now integrated in its targets.

As noted above, while direct services to humans are obviously important, they depend on adequate supporting services and these may interact with each other. It is therefore important to identify possible synergies and trade-offs among EFSs, in the short and long term, and the way in which a potential OECM would affect them. In fisheries, ESs may be more diversified and vital in coastal, densely populated areas than in the deep sea. One important action needed, therefore, is to establish an inventory of ecosystem services being used in the area in which the potential fishery-OECM is located or around it and depending on it. To the extent possible, non-fishery EFs may also be identified if they are protected or restored in the OECM.

Fishery-OECMs may be considered as a 'nature-based solution' to biodiversity degradation even though the measure may often need to be complemented by technical non-spatial measures to achieve the expected benefits. They may contribute to the protection and sustainable use of many ESs which might not all be optimized at the same time, requiring processes such as marine spatial planning to improve compatibility between services and resolve trade-offs.

Regardless of the conceptual formulation of ecosystem services, biodiversity has key roles in relation to all ecosystem services (Mace et al., 2012).

- As a regulator of ecosystem processes, it contributes to nutrient cycling, pollutants trapping, primary production, secondary production though nursery areas, feeding areas and predator–prey relationships, etc.

- As provisioning services to humans, it produces food and pharmaceuticals. It builds productive biogenic habitats which can also protect vulnerable coastlines and human settlements from erosion and severe storms (e.g. coral reefs, mangroves). It also supports recreation (e.g. scuba diving) and cultural activities, including education. Finally, it strengthens resilience and adaptation to environmental change;

- As goods, such as food, fibre, fuel and genetic resources that may be valued in economic terms but also in many other social cultural and even economic ways (IPBES, 2022b).

Part (a) of the subcriterion indicates that the maintenance of ecosystem services and their equitable sustainable use are necessary conditions of successful biodiversity conservation, and acknowledges an explicit priority on uses by people depending directly on those services, in particular Indigenous Peoples and other local communities. It also calls for accounting for trade-offs and interaction between interconnected services and functions. This is necessary because not only are different services interconnected in the social-ecological systems providing them, but use of one service by one community of users may affect the state or availability of other services to other users (or to themselves in the future). Taking these interactions into account is complex, however, and probably not possible with standard accounting tools. Different sectors of society and different cultures may attach different values to the same ecosystem services, experience changes in such services on different scales (e.g. long term vs short term; local vs global), and have different circumstances affecting their dependence on and value for biodiversity (Daw et al., 2016; Rice et al., 2025).

Consequently, although 'equity' is required in part (a) of the subcriterion, how equity is established and maintained is yet to be determined. Given the complexities of negotiating the CBD COP Decisions on the Convention Article 8j,[41] determination of how equity is judged and graded may be difficult. Thus, although D1a stresses that the end biodiversity outcome should always be positive, part (b) of the subcriterion highlights the challenges in determining what a 'positive outcome' for humanity would actually be. For consumptive uses of biodiversity like fisheries, such determinations are inherently difficult, because consumptive uses, even when sustainable, necessarily affect the biodiversity elements that are harvested, and those with tight trophodynamic linkages to the harvested species. Like all consumptive uses, ensuring fisheries are sustainable is at least a minimum necessary benchmark for the subcriterion, but not sufficient, given that even the concept of sustainability has become complex and multidimensional (see Section 2.4.3.2 and IPBES, 2022a).

2.4.4.2   Subcriterion D2: Cultural, spiritual, socioeconomic and other locally relevant values

Addressing relevant values requires that: (a) Governance and management measures identify, respect and uphold the cultural, spiritual, socioeconomic and other locally relevant values of the area, where such values exist; and (b) Governance and management measures respect and uphold the knowledge, practices and institutions that are fundamental for the in situ conservation of biodiversity.

This criterion is clear, and indicates that the establishment of OECMs should account not only for ecological values but also for the human dimensions of relevance in the area, which indeed are of importance in most if not all coastal communities and small-scale fisheries, particularly for Indigenous Peoples and other local communities. The message here is two-fold: first, OECMs (and other management and conservation regulations) cannot be implemented without due regard for the corresponding human dimensions relating to the OECM, specifically respect for and upholding of the cultural, spiritual, socioeconomic and other locally relevant values. While this might be seen as a constraint on conservation actions, the second aspect is crucial: having those values taken into account is essential, not only in its own right but also because it is an ingredient of conservation success, since the relevant 'knowledge, practices and institutions' 'are fundamental for the *in situ* conservation of biodiversity'.

Cultural values relate to people's history, identity, traditional institutions and rights. They reflect the accepted rules and behaviour of the community/society that depend on or will be affected by the OECM. In particular, an area that is already being conserved, or being considered for conservation efforts, may well have cultural value to Indigenous Peoples and other local communities – cultural value that may be based on not using that area (e.g. in the case of a tabu) or on sustainably using the area (e.g. as a source for a culturally important harvest).

Spiritual values relate to religious, moral or ethical beliefs, specifically related in this case to the relations between humans and nature and more specifically to biodiversity and its conservation. Many Indigenous Peoples have such values underlying their sustainable use of places and nature, e.g. Netukulimk, in the case of the Mi'kmaw people in eastern Canada (McMillan and Prosper, 2016).

Socioeconomic values are best viewed in a broad sense, as an amalgamation of social and economic values, and those that are intermediate. Economic values, including aspects such as income and exports, have long been recognized as important in commercially oriented fisheries, with well-established methods for their collection and analysis. Intermediate values may be seen as economic values with an added human dimension, e.g. relating to employment, labour migration and work conditions. Social values, such as social cohesion, community stability, conflict resolution and power structures, are not specifically mentioned in the Decision but are incredibly important in general, and specifically relevant to the success of spatial management measures (e.g. Charles and Wilson, 2009; Christie *et al.*, 2017; Westlund *et al.*, 2017; Quimby and Levine, 2018).

Other locally relevant values can also play important roles. One value that often arises is a 'sense of place' – a strong feeling of those in a specific location, e.g. a coastal community, of caring for that 'place'. The sense of place (which could also be seen as a social value, that may or may not be cultural or spiritual) leads in many cases to a particular local concern for the local environment, resulting in effective conservation stewardship (Charles, 2021).

Cultural, spiritual and other locally relevant values may be particularly important in coastal areas and small-scale fisheries, particularly for Indigenous Peoples and other local communities. In all such areas and fisheries, people are interacting closely with the ocean and the biodiversity within it. This is also where there may most likely be locally managed ocean areas, sacred sites and so on. These values may not be easy to quantify in individual cases (IPBES, 2022b) but they are real and generate consequences. They may decrease in importance from the coast to the open ocean and deep sea, but there is increasing acknowledgement that many cultures attach significant value to the ocean and its biodiversity, including existence values even for those living far from the ocean (e.g. Costanza, 1999; Rock et al., 2020; Whitty, 2023; Perez-Alvaro and Boswell, 2025). The highlighting of these values in Decision 14/8 draws attention to the wide diversity of human dimensions of biodiversity conservation that can be considered in addition to the more material values of food, revenues, recreation, livelihoods and biodiversity values usually accounted for in the conventional fisheries sustainability objectives of OECMs.

Notes

[1] As defined in the Convention itself (§2).

[2] Considering the lack of enforcement mechanism, the obligation represented by the definition may be considered, at least, as a moral obligation.

[3] www.un.org/bbnjagreement/en/bbnj-agreement/text-bbnj-agreement (accessed 17 July 2025).

[4] www.protectedplanet.net/en/resources/wdpa-manual (accessed 17 July 2025).

[5] However, Decision 14/8 strongly suggests that, following the acqusition of the OECM status, the co-benefits that justified that status be identified in the OECM objectives, becoming intended benefits, inter alia to guide performance assessment. It also requires that sustainable use and conservation be consistent.

[6] Negative impact of other sectors on the same biodiversity of concern would still be expected when evaluating the candidate OECM against the CBD criteria.

[7] Although PA effectiveness has been a long-standing concern and strong efforts are being made to improve it, e.g. through the IUCN's Green List of Protected and Conserved Areas, a standard for assessing and recognizing protected areas that demonstrate effective, equitable and successful conservation outcomes.

[8] In this book, the term 'status' will be used.

[9] For example, North Sea Sandeel Fishery closure; Rockall/NEAFC Haddock Box; Lophelia pertusa Coral Conservation Area in Canada; the NAGE Sponge VME, in NAFO.

[10] For example, ABFM, community-based conservation area, LMMA, or refuge.

[11] Fishery restriction measures are considered as potential OECMs in the European Union Biodiversity Strategy for 2030. Under the Strategy, OECMs will be taken into account if they provide some form of legal protection that indirectly promotes conservation.

[12] Annex III, A on guiding principles and common characteristics concerning the identification of OECMs.

[13] In CBD Decision IX/20 (2008) and Annexes I and II. See also www.cbd.int/ebsa/ (accessed 17 July 2025).

[14] Fisheries and Oceans Canada, Bedford Institute of Oceanography, Dartmouth, Nova Scotia, Canada.

[15] Executive Secretary, Northwest Atlantic Fisheries Organization (NAFO), Halifax, Nova Scotia, Canada.

[16] With all the controversies this process raised, e.g.in Mrozowski, 1999; Mace, 2014; Bermejo and Bermejo, 2014; Feydel and Bonneuil, 2015; Smessaert et al., 2020.; Standing, 2024.

[17] www.cbd.int/traditional/intro.shtml, www.cbd.int/abs/infokit/revised/web/factsheet-tk-en.pdf (both accessed 17 July 2025).

[18] Fishery-related threats relate to overfishing, by-catch or destructive fishing practices on vulnerable habitats, leading *inter alia* to excessive mortality, lower spawning potential, changes in age, species composition and trophic web, and lower resilience to climate oscillations and change. Fishery threats may or may not be as serious as threats from other economic marine or land-based sectors or from changes in environmental conditions. Further, there may be compounding of the threats from multiple sources.

[19] The risk being, for example, the likely cost of the impact of a potential threat multiplied by the probability that such threat really occurs.

[20] www.cbd.int/decision/cop?id=11,663 (accessed 17 July 2025).

[21] Because of threats from fisheries or other sources of impact.

[22] Rules that trigger preagreed action in foreseen situations to avoid damaging delays in responses.

[23] With noticeable exceptions in a few advanced countries.

[24] TURFs allocate exclusive harvesting rights for one or more marine species in a specified geographical area. TURFs are ideal for species like abalone that will not move beyond TURF boundaries, but they can be designed for more mobile species as well. TURFs may occur independently or they may be part of a broader system of TURFs. Well-designed networks of TURFs can be used to manage more complex fisheries, including those with mobile species and multiple groups of fishermen.

[25] www.un.org/development/desa/indigenouspeoples/wp-content/uploads/sites/19/2018/11/UNDRIP_E_web.pdf (accessed 18 July 2025).

[26] www.ihrb.org/resources/what-is-free-prior-and-informed-consent-fpic (accessed 18 July 2025).

[27] For example, in the core Decision (§5b) as well as in Annex II (§6, 7e, 11a, 11 hour, 12d,), Annex III (§C1f) and Annex IV (§A1i, C1c, C2b, C3, D6h).

[28] This was confirmed in various technical meetings involving the CBD Secretariat.

[29] http://wcmc.io/WDPA_Manual (accessed 18 July 2025).

[30] In some cases, the seasonal closure involves the entire EEZ and the fleet is locked in port but this situation is not considered relevant here.

[31] Including CBD Decision 14/8 (2018), which includes in its Annex II the Voluntary Guidance on Effective Governance Models for Management of Protected Areas, including Equity, Taking Into Account Work Being Undertaken under Article 8(J) and Related Provisions. See also Khotari *et al.* (2012).

[32] Regional fishery bodies (RFBs) are intergovernmental bodies through which States co-operate on the management of fisheries in specific regions. Many RFBs, particularly those established under FAO (e.g.CECAF, WECAFC, etc.), deal with fishery research and capacity building but have only an advisory role in management and their competence area may cover both areas under national jurisdiction and in the high sea. RFBs establishing binding measures for conservation and sustainable management of highly migratory or straddling fish species, in the high sea, are referred to as regional fisheries management organizations (RFMOs). Their membership includes all countries with fishing interests in the area. A regional fisheries management arrangement (RFMA) is any form of arrangement through which States adopt conservation and management measures that does not provide for the establishment of an organization. www.fao.org/in-action/vulnerable-marine-ecosystems/background/regional-fishery-bodies/zh/ (accessed 18 July 2025).

[33] Regional Seas Conventions are agreements among countries sharing a common body of water, setting rules for the management of marine resources in that region. State Parties adopt a legally binding convention setting out what governments must do to implement their action plan. Most conventions have additional separate protocols addressing specific issues – such as protected areas or land-based pollution – in more detail. www.unep.org/topics/ocean-seas-and-coasts/regional-seas-programme (accessed 18 July 2025).

[34] For example, in SE Australian fisheries, spatial closures have seen very small changes over the years. Most remained and it can be expected that they will be either permanent or change very slowly, despite the lack of legal certainty, because (i) they are seen as contributing to sustainability; (ii) they are not removed even after rebuilding; and (iii) in multispecies fisheries, not all species recover at the same pace. Shifts of some closures in response to climatic oscillations and change have been considered but are not always accepted by all stakeholders (Keith Sainsbury, personal communication).

[35] In fisheries, often the target species of the fishery and features of its habitat or prey spectrum. In conservation, protected, endangered and threatened species.

[36] Such as non-essential habitats, non-target species in fisheries. Species and habitats raising no concern in conservation.

[37] International Guidelines for the Management of Deep-sea Fisheries in the High Seas (Rome, FAO, 2009).

[38] The Decision refers mainly to Aichi Target 11, which is not specified in this respect.

[39] www.iucn.org/theme/nature-based-solutions (accessed 18 July 2025).

[40] www.nature.org/en-us/what-we-do/our-insights/perspectives/three-things-nature-based-solutions-agriculture/ (accessed 18 July 2025).

[41] CBD Article 8j states: *Subject to national legislation, respect, preserve and maintain knowledge, innovations and practices of indigenous and local communities embodying traditional lifestyles relevant for the conservation and sustainable use of biological diversity and promote their wider application with the approval and involvement of the holders of such knowledge, innovations and practices and encourage the equitable sharing of the benefits arising from the utilization of such knowledge innovations and practices.*

3 OECM Implementation Cycle

Abstract

This chapter describes the full implementation cycle of OECMs from their initial identification to their final reporting at the international level. The cycle is divided into phases. The identification phase is subdivided into the steps needed to (i) consolidate the information available; (ii) undertake a first quick screening of ABFMs, to identify potential fishery-OECMs; (iii) undertake a full assessment of the most promising ones; and (iv) informing the legitimate authority for its decision regarding the eventual fishery-OECM status. The chapter also stresses the importance, for effective implementation of this cycle, of adequate enabling frameworks, equitable governance and active participation of stakeholders, and the need to consider the best available science and local knowledge, and all important locally relevant values. A specific section is dedicated to the need for capacity building.

This chapter and indeed the whole book illustrate how an OECM implementation cycle could be organized to fully and effectively implement Decision 14/8 and contribute to the goal of biodiversity conservation. In the text, the term 'needs' is often used to indicate something necessary for the cycle to comply with the intent of Decision 14/8. Thus, unless specified otherwise, 'needs' refer to actions or steps that are 'necessary' to complete the proposed cycle, possibly in a variety of ways, but not mandatory.

Decision 14/8 contains important guidance on OECMs, their identification and governance, management and reporting. Figure 3.1 is a graphic summary of that guidance, applicable whether the potential OECM is an existing ABFM or newly created with the intention of becoming an OECM. The sequence of actions needed could be logically organized into five phases.

1. *Stepwise identification*, including data consolidation on information on ABFMs, quick screening to detect potential OECMs, full assessment of potential OECMs, and provision of advice on candidate OECMs and potentially upgradable OECMs.
2. *Decision making* by the legitimate authority in a participative process, formally identifying which ABFMs are given OECM status and which may be allocated the resources needed to maintain that status. Positive

decisions can be reviewed whenever the performance of the OECM is deemed inadequate (e.g. below the minimum requirements of the Decision or the national standards).
3. *Reporting* at international level, to the WCMC database on the identification of OECMs and, later, on their performance.
4. *Management* of the OECMs as special ABFMs requiring specific regulations and monitoring to deliver and document the expected outcomes.
5. *Monitoring and evaluation* of the OECM's performance, necessary to maintain OECM status in the long term. The periodicity of the evaluations must take into account the life histories of target species and other key species that were influential in the initial OECM assessment, and the rate of progress expected towards the desired outcomes.
6. *Conclusions on performance* and related advice to be provided to the legitimate authority after recurrent assessment.

Auditing is not mentioned in Decision 14/8 but is becoming part of best practice in management to check on compliance with legal texts and procedures as well as cost vs benefits. Auditing may take different forms depending on the kind of OECM or fishery operating in a given location; for example, a locally managed marine area (LMMA) with a small-scale fishery may

© Serge M. Garcia, Jake Rice, Anthony Charles and Daniela Diz 2025. *Other Effective Area-based Conservation Measures (OECM) in Marine Capture Fisheries: Identification, use and performance assessment* (Serge M. Garcia *et al.*)
DOI: 10.1079/9781836990888.0003

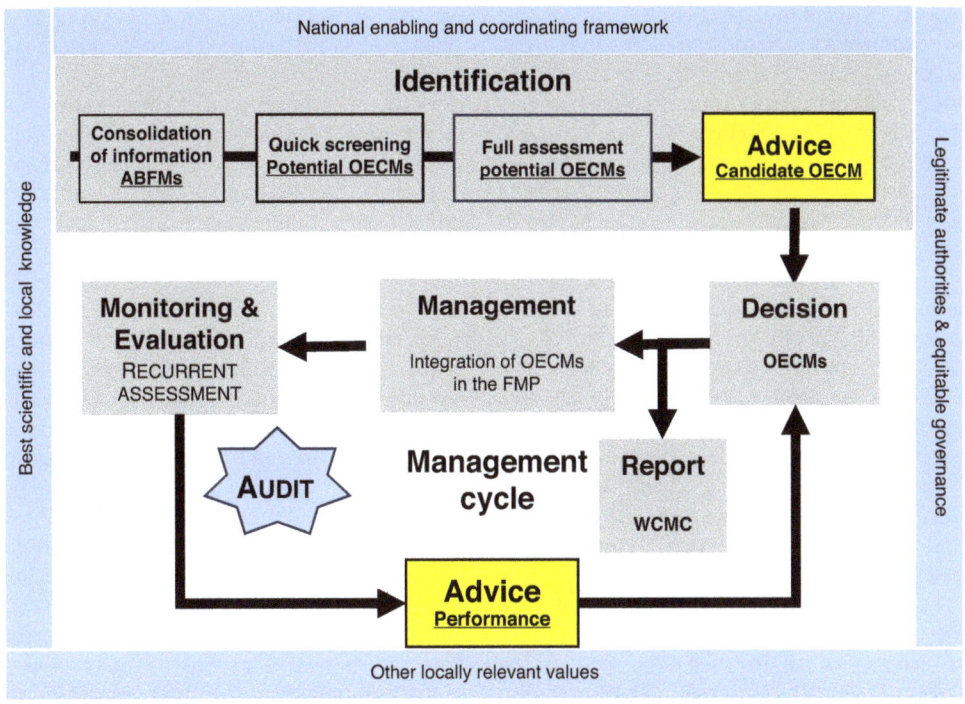

Fig. 3.1. Suggested fishery-OECM implementation process (modified from Garcia *et al.*, 2019). Figure used with permission.

have a different process from a governmental OECM in an industrial fishery. In any case, it should be a part of a well-managed fishery, regardless of whether it uses OECMs or not. However, the factors that were important to the assessment recommending OECM status might well be factors that should be reviewed in each audit, as they might change over time, together with costs and benefits. Similarly, the procedures necessary to maintain OECM status might require changes over time and warrant review as part of the audit.

This logical cycle is not imposed or explicitly proposed in the Decision. However, it emerges progressively and naturally as the steps needed to first make and then implement the OECM Decision are integrated with the activities already needed for effective fisheries management. Each State may organize its process as it wishes, taking into account its own circumstances, but there are some logical aspects of this cycle, such as assessment only being possible after information has been assembled and quality assured,

and registering the OECM with the WCMC only being possible after a positive decision on OECM status and international reporting.

Thinking about this implementation cycle in a fisheries context is important because a fishery-OECM has to be part of any existing fishery management plan and cycle of the fishery in which it operates. This is the case for new areas created as OECMs from the onset, as well as for existing ABFMs identified as OECMs that have still to contribute to the fisheries objectives while providing positive biodiversity outcomes. The classic adaptive management cycle used for fisheries and conservation management, both formal and informal,[1] involves (i) determination of objectives; (ii) identification of corresponding indicators and reference values; (iii) decisions on the management strategy, plan and measures; (iv) implementation and enforcement; (v) monitoring of the fishery and the resources; (vi) performance evaluation and, based on its results, (vii) adaptive adjustment of the management system or the policy itself, as necessary, to achieve or adapt the objectives if that is needed to result in a sustainable fishery. The implementation cycles of the fishery-OECM and that of the fishery within which it occurs are similar and, in practice, would be synchronized and integrated as much as possible, to avoid duplication of steps and ensure consistency of the products of both cycles. The integration process is addressed in detail in Chapter 6.

Figure 3.1 also illustrates, with banners around the picture, the fact that, in order to comply with the CBD Decision 14/8 the effective use of fishery-OECMs requires: (i) a legitimate authority, legitimized in law and policy, and recognized as such by its constituency; (ii) an equitable governance process (as defined in CBD Annex II, B); (iii) use of the best scientific and local knowledge available; and (iv) recognition and accounting of ecological, economic, social and other locally relevant values.

The implementation process requires financial, technical and human resources to accomplish all the tasks foreseen in the different phases of the cycle. The process also depends on involvement of key institutions of the fishery management system –such as the statistical service (for fishery data) or the Coastguards (for compliance data). Linkages with these other agencies are not addressed in this book, but are assumed to be part of any fishery well enough managed to have areas that warrant OECM status.[2] For these reasons, the process also calls for collaborations with conservation agencies, for specialized information and technical and scientific competences. Considering the usual limitations of research and management budgets, collaborations with other institutions are valuable to pool competences, reduce potential duplications and conflicts, facilitate synergies among agencies, area-based measures, and spatial and non-spatial measures, and enhance cross-sectoral collaborations.

Ways in which these tasks can be conducted are addressed in Chapters 5–8.

3.1   Capacity Building

Capacity building is addressed in practically all international agreements, particularly but not only to ensure that developing countries, including Least Developed Countries (LDCs) and Small Island Developing States[3] (SIDS), and low-capacity communities, including Indigenous Peoples and local communities, develop their implementation capacity, including through financing, co-operation, technical assistance, transfer of technology, etc. In this book, this issue is referred to for many phases of the OECM implementation cycle to point out the specific potential needs for that phase. However, these needs are regrouped in Table 3.1 to provide States and managers with a blueprint of their potential needs, to allow them to (i) consider what capacity they already have and can autonomously mobilize; and (ii) identify the aspects for which they may need external co-operation or assistance.

The comprehensive inventory of actions given in Table 3.1 may look overwhelming but the specific capacity needed for many actions may already be available, for example in the existing fisheries management system, or may be easily complemented. The specific needs for each action are therefore case and context specific and can only be determined locally in collaboration with the stakeholders involved.

Highly participatory and flexible pilot initiatives, adapted to the recipients' contexts and co-driven with them, are more likely to prioritize and optimize the use of intrinsic and intrinsic capacity-building efforts, enhancing the sense of ownership of the process and its outcomes.

Capacity needs in areas with limited data and competence are likely to be more crucial than those for centralized States or ministries, and appropriate ways need to be found to ensure that new competences and assessment burdens can be effectively absorbed.

A pragmatic approach to capacity building is through 'learning by doing', experimenting with pilot initiatives and progressively scaling up from there. A 'comprehensive approach' to identification, for example, in which all potential OECMs are addressed simultaneously, is more demanding than an 'incremental approach' in which the process starts with one or a few pilot areas and proceeds as experience and capacity improve. Finding how to better mobilize, use and enhance available capacity, in feasible increments, may often be a wise long-term capacity-building strategy,

Table 3.1. Actions that may require capacity building for the different processes in the fishery-OECM implementation cycle: enabling, governance, management, assessment (for identification and recurrent evaluation). It is assumed that the reporting capacity already exists. For each 'action' the relevant criteria or subcriteria are in brackets.[4]

Process	Selected actions potentially requiring additional capacity
Enabling process (B2, B3)	• Update fisheries and conservation acts and policies (B2, B3) • Strengthen inter-ministerial co-ordination and cross-collaborations (MSP) (B2) • Provide incentives including financing of management for fishery-OECMs (B3)

Governance (B2, C2)	• Clarify local and central legitimate authorities/responsibilities (B2)
	• Identify/strengthen local governance (incl. IPs and local communities) as appropriate (B2)
	  o *Enhance local collaboration frameworks (B2)*
	  o *Identify/involve stakeholders; enhance participatory processes (B2)*
	  o *Seek equity: recognition, participation, equitable distribution (B2)*
	  o *Clear statement of 'long-term' intent in policy and management (C2)*

Management (B3, C1, C2, D1)	• Update biodiversity-related objectives, targets, indicators (B3) • Update biodiversity-related enforcement (B3) • Develop risk-assessment and management capacity (B3; C1) • Determine the 'sustained' management intent (C2) • Strengthen adaptive management capacity (projections, decision rules) (B3, C1) • Strengthen EAF (B3)l • Integration of OECMs in the FMP (B3, C1) • Protect particularly 'supporting' ecosystem services (D1)
Assessment for identification and evaluation (B1, B3, C2, C3, C4, D1, D2)	• Mobilize best scientific evidence and local knowledge (C2, C3, D2) • Enhance multidisciplinary assessment capacity (B3, C1, D2) • Strengthen stakeholders' analysis (B3) • Develop/access GIS capacity (B1) • Analysis of ecosystem services, their interactions and trade-offs (D1) • Identify locally relevant values (D2) • Document use of local institutions, practices, local knowledge (A, C4, D1) • Identify biodiversity values (i) in the general area, and (ii) in the OECM (C3) • Identify specific attributes of concern from fishing activities (C3)
Monitoring (C4)	• Improve formal or informal monitoring systems (C4) • National ABFM registry (e.g. for a comprehensive identification approach (B1, C4) • National registry of MPAs and OECMs (A, C4) • Accessible archiving system for OECM information (C4)

Used with permission from CBD (2023a).

considering the potentially large number of ABFMs that can be considered as potential fishery-OECMs.

In closing this section, it is important to stress that the capacity-building needs for fishery OECMs may be usefully nested in other national and regional needs already identified and being implemented, e.g. for education, livelihoods or blue economy development.

Notes

[1] In informal management systems, the elements (i) to (vi) will usually not be available in formally written and archived texts, but may be collected through questionnaires and cooperation with local knowledge holders.

[2] Some small-scale fisheries do not have standard engagement with statistical offices, Coastguard or other enforcement agencies, etc. They can still receive OECM status if the community and cultural practices used for evaluating resource status and ensuring compliance with the practices of the community management are sufficient to meet the OECM standards for Indigenous Peoples and other local community settings.

[3] Often referred to now as Large Ocean States (LOS).

[4] Table 3.1 was developed for a CBD-FAO-FEG meeting on regional capacity building for fishery-OECMs, 17-19 May 2023.

4 Enabling Frameworks

Abstract

Legal, policy and financial overarching enabling frameworks should guide and facilitate the governance and management of fisheries implementation and, more specifically, the identification, management, evaluation and reporting of OECMs. This chapter briefly considers the frameworks available at global level, through the United Nations, its general assembly, its conventions (like UNCLOS and the CBD), its agencies (like FAO) and programmes (like UNEP) and its international agreements (like UNFSA and the BBNJ). At the regional level, RFMOs and RSCs play a fundamental role, co-ordinating States' roles in international jurisdictions. The frameworks available at national level, and particularly the fishery and legal frameworks, are considered in more detail, emphasizing the actions needed to facilitate the OECM identification and integration processes. The tensions regarding the compatibility of industrial fisheries and OECMs are examined.

The implementation of OECMs is facilitated and indeed enabled by the overarching policy, legal and financial, and collaborative frameworks necessary to ensure, and improve as appropriate, the overall political will, legal support, financial back-up and cross-sectoral and institutional collaboration between conservation and sectoral authorities. These enabling factors are important in determining the quality of governance. In many cases, the existing governance frameworks may already be adequate. In others, capacity building might be required for both cross-sectoral and sectoral OECMs, to ensure the needed empowerment, co-ordination, effectiveness, efficiency and consistency in their identification and management. The enabling needs are referred to in Decision 14/8, in relation to the policy and finance environments (Annex I, II, B, e, f), diverse and equitable governance (Annex I, II, §9, §12), effective measures for *in situ* biodiversity conservation (Annex III, C, 1, e), and capacity building (Core text, §10).

At the global level, the overarching legal and policy framework is provided by the 1982 United Nations Convention on the Law of the Sea (UNCLOS), the 1992 CBD, the 1995 United Nations Fish Stocks Agreement (UNFSA) and the recently adopted Agreement on the Conservation and Sustainable Use of Marine Biological Diversity of Areas Beyond National Jurisdiction (in short, the BBNJ Agreement[1]). The BBNJ Agreement has implications for the implementation of area-based measures in general and OECMs in particular, in the high seas and the area.[2] Although it is too early to foresee clearly the future interactions between CBD Decision 14/8 and the BBNJ Agreement, it should be noted that both the CBD and the BBNJ Agreement: (i) aim to uphold sustainable use and conservation of biodiversity in their areas of competence; (ii) have capacity building as a priority; (iii) encourage the use of area-based management tools; and (iv) stress the importance of scientific research and data sharing. Clearly, synergy between the CBD Global Biodiversity Framework (GBF) and the BBNJ Agreement framework (particularly in relation to GBF Target 3) is needed for greater efficiencies in the implementation of their aligned goals. Opportunities for collaboration and co-ordination exist and will increase as soon as the BBNJ Agreement enters into force. The CBD Decision 14/8 established the international framework enabling the identification and implementation of OECMs in all ecosystems (terrestrial and marine) and economic sectors, referring briefly to the policy and finance enabling frameworks (p. 5) and the enabling conditions needed to mainstream biodiversity conservation in economic sectors and improve equity (p. 8).

© Serge M. Garcia, Jake Rice, Anthony Charles and Daniela Diz 2025. *Other Effective Area-based Conservation Measures (OECM) in Marine Capture Fisheries: Identification, use and performance assessment* (Serge M. Garcia *et al.*)
DOI: 10.1079/9781836990888.0004

At the regional level, there already exist international arrangements to manage transboundary and high seas resources, including through ABFMs, in line with the 1995 UNFSA. In some cases, minimal updates of the arrangements may be needed to identify and manage OECMs and the specific biodiversity attributes. All regional fishery bodies (RFBs) and regional fisheries management organizations (RFMOs) are slowly progressing in effective implementation of EAF in which areas granted OECM status (particularly VMEs) may have an important role to play. These regional organizations can therefore be mandated by their State Parties to identify OECMs with little or no need to further update their conventions and management frameworks. This is confirmed by the ongoing efforts in NAFO, NEAFC[3] and GFCM to identify fishery-OECMs in their regulatory areas. Following a robust science-based assessment of some VMEs and the effectiveness of the respective management measures against the CBD OECM criteria, NAFO and NEAFC identified a number of protected VMEs as areas that deserve OECM status and have registered them in the WCMC WD-OECM database (see Boxes 2.1, 4.1 and 4.2). The potential for identifying OECMs in tuna-RFMOs and large oceanic pelagic fisheries is still unclear.

The necessary collaboration between RFBs including RFMOs and Regional Seas Conventions is slowly increasing, particularly in the north-east Atlantic and the Mediterranean, and actively promoted by the CBD through the Sustainable Oceans Initiative (SOI).[6] NEAFC and OSPAR, for example, have harmonized their respective MPAs and OECMs to ensure alignment with CBD Decision 14/8 and avoid any spatial overlap (D. Campbell, personal communication and, therefore, any double-counting towards GBF Target 3. In the case of the NE Atlantic, the collective arrangement between competent international organizations on co-operation and co-ordination regarding selected areas in ABNJ provides a platform for co-operation, including on MPAs and OECMs, especially between OSPAR and NEAFC and increasingly other competent organizations.[7]

The General Fisheries Commission for the Mediterranean (GFCM), for example, established in 2005 11 fisheries restricted areas (FRAs) to protect vulnerable marine ecosystems (VMEs

FRAs) and essential fish habitats (EFHs FRAs) against significant adverse impacts (SAIs) from fisheries and to enhance resources productivity. In 2022, GFCM FRAs were identified as good potential OECMs, and GFCM and FAO started assisting GFCM members in assessing their FRAs against OECM criteria. One of the FRAs is a general ban on bottom fishing in the whole Mediterranean below 1000m (Box 4.2).

The following sections refer briefly to the overarching national governance framework within which the use of OECMs is necessarily nested, before focusing on fishery governance and legal frameworks.

4.1 National Governance Framework

This section does not intend to analyse in any detail the national framework eventually put in place or strengthened to deal effectively with OECMs across all sectors and jurisdictions. However, the elements of importance for mainstreaming fishery-OECMs are briefly summarized below.

Governance has been defined in many ways. A synthetic definition is:

> A systemic concept relating to the exercise of economic, political, and administrative authority. It encompasses: (i) the guiding principles and goals of the sector, both conceptual and operational; (ii) the ways and means of organization and co-ordination of the action; (iii) the infrastructure of socio-political, economic and legal instruments; (iv) the nature and *modus operandi* of the processes; and (v) the policies, plans and measures. (Garcia, 2009)

The 'governance' term is often used to cover two interconnected and partially overlapping levels of administration (e.g. in Kooiman, 2005; Charles, 2023): (i) at the strategic level, the institutions, processes, policies, strategies, laws, overarching rules and oversight; and (ii) at the operational level, the regulations, implementation means, monitoring control and surveillance (MCS), and performance assessment. This operational level is usually referred to as 'management' and distinguishes from policy and planning. The definition applies across

Box 4.1. North-East Atlantic Fisheries Commission (NEAFC) OECMs

Diz, D.[4] and Campbell, D.[5]

From 2020, different NEAFC bodies considered various area-based fisheries measures that might meet the CBD criteria for OECMs. It was agreed that the measures related to bottom fishing closures and restrictions to protect vulnerable marine ecosystems were the most closely aligned with the CBD criteria. Accordingly, a request for scientific advice was formulated to ICES on the long-term benefit of such measures. NEAFC's 2023 annual meeting received the advice from ICES (ICES, 2023a) that the closed and restricted bottom fishing areas in the Regulatory Areas achieved *in situ* biodiversity/ecosystem benefits, as long as no bottom fishing activities occurred (NEAFC, 2023a). ICES had comprehensively looked at the biodiversity attributes of those areas, including by making use of EBSA and other available scientific information. ICES confirmed that the Regulatory Area contained biodiversity attributes as articulated in the CBD guidance. It also noted that the any reopening of restricted bottom fishing areas to exploratory fishing might preclude them from satisfying sustained governance for the long term.

Accordingly, the annual meeting agreed to identify an OECM, based on measures that protected VMEs, for notification to the CBD in 2024 (NEAFC, 2023b). There was agreement to report the closed areas for protection of VME as OECMs to WCMC, where these appear as 20 different files. This would be reported when a complementary narrative on the interaction between NEAFC OECM and OSPAR MPAs, jointly developed with OSPAR, was ready (NEAFC, 2023c).

On the areas restricted to bottom fishing, a conservative approach had been applied, which focuses on depths that are shallow enough (i.e. less than 1400 m) to reflect a realistic prospect of bottom fishing impacting them. The NEAFC Annual Meeting requested ICES to provide further advice based on the polygons identified as shallower than the 1400m isobar which would open a way to further discussions at the Permanent Committee on Management and Science (PECMAS) on how to proceed. This advice would confirm (or not) that the relevant polygons contained the biodiversity attributes that would meet the OECM criteria (NEAFC, 2023c). The ICES advice concluded that each of the 14 polygons identified by NEAFC had multiple biodiversity attributes, and that therefore the ICES 2023 advice on OECMs applied to each of the 14 polygons as long as no bottom fishing activities were authorized, noting that 'any bottom fishing in the polygons may preclude them from satisfying sustained governance for long-term biodiversity benefits' (NEAFC, 2023a). NEAFC's OECM submission excludes the areas of overlap with OSPAR MPAs to avoid double counting and in adherence to CBD Criterion A (Fig. 4.1).

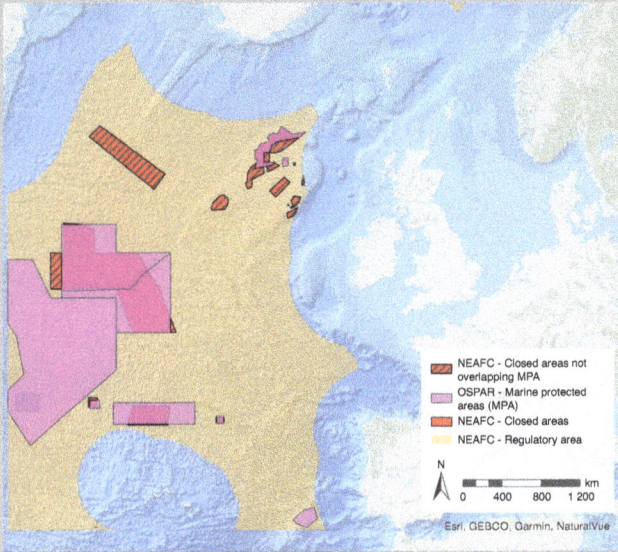

Fig. 4.1. NEAFC closed areas and OSPAR MPAs. Esri. GEBCO. Garmin. Natural Vue. Courtesy of Katharina Jacob, Fisheries Directorate, Norway. Used with permission.

Box 4.2. General Fisheries Commission for the Mediterranean (GFCM) work on OECMs

Morello, E.B. and Nastasi, A.[8]

Since 2019, the General Fisheries Commission for the Mediterranean (GFCM) of the Food and Agriculture Organization of the United Nations (FAO) has been working on OECMs in the Mediterranean and the Black Sea to assist member countries in identifying potential fisheries-related OECMs to contribute to global area-based biodiversity targets (Fig. 4.2).

Within the GFCM advisory process, OECMs were first discussed at the 2019 Working Group on Marine Protected Areas (WGMPA), when the idea of identifying areas where efficient spatial management of fisheries resulted in the highest effective biodiversity conservation outcomes was put forward for the first time. The GFCM fisheries restricted areas (FRAs; Fig. 4.3) represent the primary example of such areas, because they are managed as sustainable fisheries and can host VME indicator species. As a result, the 44th GFCM session encouraged its contracting parties and co-operating non-contracting parties (CPCs) to participate in the ongoing international process of defining and identifying OECMs, including by organizing Mediterranean-specific expert meetings in collaboration with FAO and relevant partners (FAO, 2022a). With this in mind, the GFCM Secretariat participated in several international events and partnered with the FAO Fisheries and Aquaculture Division (NFI) to explore the potential for identifying fisheries-OECMs specifically in the Mediterranean Sea. In late 2021 and early 2022, the FAO-GFCM events allowed the rolling out of a process through which the OECM concept was introduced to the GFCM area of application and case studies were identified for their potential evaluation against the criteria for OECMs defined by the CBD.

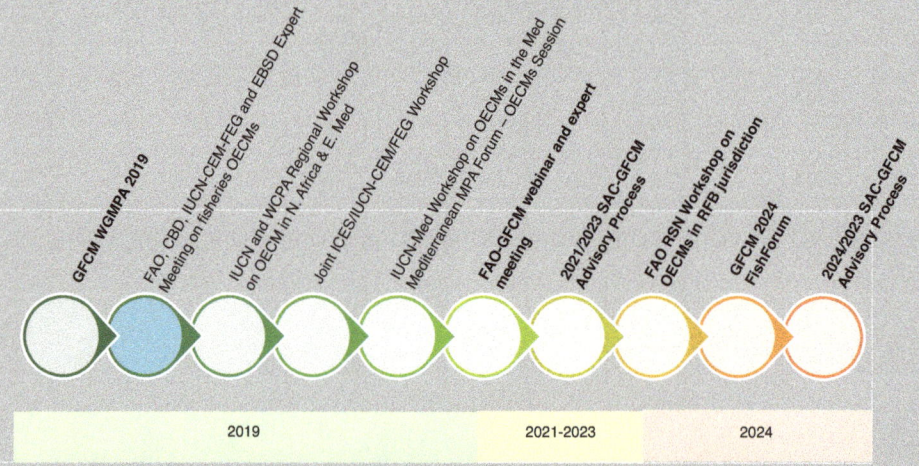

Fig. 4.2. Annotated timeline of GFCM work on OECMs. Used with permission from the FAO-GFCM.

A FAO-GFCM webinar and an expert meeting on fisheries-related OECMs were organized, resulting in the initial screening of all FRAs and of some additional case studies against the simplified criteria retained by FAO (FAO, 2022, 2023). Remarkably, these meetings were attended not only by fisheries and biodiversity experts and scientists, but also by representatives of CPCs from ministries of agriculture/fisheries and of environment/ecological transition. Notably, several FRAs stood out as potential OECMs and were further discussed in 2022 at the four GFCM Subregional Committees (SRCs), which cover the western, central and eastern Mediterranean and the Adriatic Sea.

Continued

Box 4.2. Continued

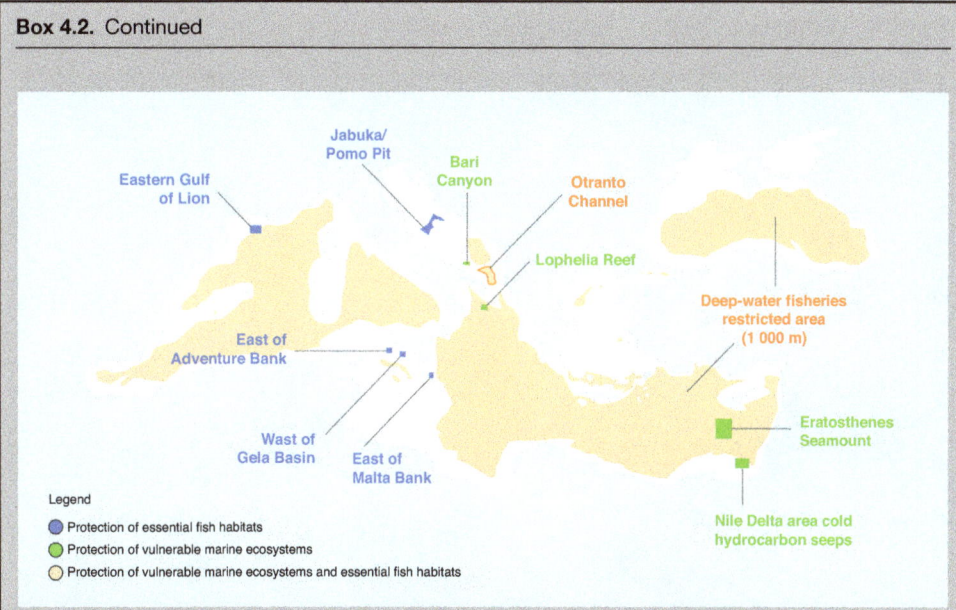

Fig. 4.3. Fisheries restricted areas established by the GFCM between 2005 and 2024. Used with permission from the FAO-GFCM.

Considering the data and evidence available, in 2022 three FRAs were initially assessed as potential OECMs worthy of full assessment: the East of Adventure Bank and Gela Basin FRAs, established in 2016 in the Strait of Sicily, and the Jabuka/Pomo Pit FRA, established in 2019 in the Adriatic Sea (see Fig. 4.3). These FRAs were established by the GFCM to enhance the productivity of essential fish habitats of demersal stocks such as European hake (*Merluccius merluccius*), Norway lobster (*Nephrops norvegicus*) and deep-water rose shrimp (*Parapenaeus longirostris*) while also protecting the known presence of VME indicator species. These areas are permanently closed to demersal fishing and, in the case of the Jabuka/Pomo Pit, also to pelagic fishing. The analysis of available data has provided evidence of their effectiveness, in terms of both compliance (i.e. suspended fishing activity) and increased biomass of demersal stocks. In addition to these areas, other FRAs addressing the protection of sensitive benthic habitats (i.e. VME closures) could also qualify as potential OECMs, but no screening exercise has been initiated as of the time of writing. An additional FRA has been established since the 2022 screening exercise: the Otranto Channel FRA (southern Adriatic Sea) was adopted in November 2024 (see Fig. 4.3), addressing both essential fish habitats for commercial crustacean species and VMEs formed by the deep-sea soft coral *Isidella elongata* (Fig. 4.4). This FRA was not assessed in 2022 but could represent a promising addition to the pool of potential fisheries OECMs in the Mediterranean.

Continued

Box 4.2. Continued

Fig. 4.4. *Isidella elongata* in the Otranto channel . ©FAO/GFCM. Used with permission from the FAO-GFCM.

However, the GFCM Scientific Advisory Committee on Fisheries (SAC), and subsequently the annual session of the GFCM, advised that the GFCM Secretariat should provide support to CPCs, upon request, while further activities would be contingent on the outcomes of future discussions among CPCs. In 2023, the discussion resumed with a new roadmap towards progressing on a full assessment of FRAs against the CBD criteria. In response to this request, in 2023 the SAC and the Commission raised a few procedural questions on the OECM process to be addressed by the GFCM Secretariat, in collaboration with FAO, the United Nations Environment Programme (UNEP) and the CBD Secretariat, in order to present a comprehensive portfolio of views at the 47th annual session of the GFCM in November 2024.

In January 2024, the GFCM Secretariat, along with many other regional fishery body (RFB) secretariats, participated in a workshop on OECMs under the jurisdiction of RFBs – including regional fisheries management organizations (RFMOs) – organized by FAO NFI and the Regional Fishery Body Secretariats Network (RSN). Many of the same procedural questions and challenges to the OECM process were raised, shared and discussed in depth. In particular, the implications and opportunities arising from the identification of fisheries OECMs in areas under the jurisdiction of RFMOs were assessed, and several case studies relevant to RFBs were reviewed. The exchange with other RFBs proved extremely valuable, and the outcomes were presented to the GFCM Fish Forum in 2024 (19–23 February 2024, Antalya, Türkiye), as well as to the GFCM SRCs and the SAC in 2024. To describe the main outcomes of the process and summarize the steps taken towards the OECM identification and submission process, the GFCM Secretariat presented a working document (GFCM, 2024) to the 47th session of the GFCM (November 2024). This document was intended to provide CPCs with the main elements needed for assessing the implications and opportunities related to the identification and recognition of fishery-related OECMs in the Mediterranean and the Black Sea. Importantly, it (i) reinforced the fact that embracing OECMs in the fisheries domain is not about introducing new conservation obligations, but rather about recognizing and valuing the additional positive outcomes

Continued

Box 4.2. Continued

that effective fisheries management is already delivering in terms of supporting biodiversity conservation, and (ii) identified the possibility of giving a mandate to the GFCM Secretariat to report FRAs as
OECMs on behalf of CPCs.

GFCM working groups have expressed doubts on the possibility of designating the entire 1000m
FRA as a fishery-OECM for various operational reasons and suggested that a number of smaller, more
manageable areas might provide better potential fishery-OECMs. For instance, the identification of
biodiversity attributes, fishable depths and sustained biodiversity outcomes[9] could help refine these
large FRAs to the purpose and spirit of what the global target on OECM is trying to achieve. FRAs
granted OECM status would be established by binding decisions, including within the GFCM multiannual management plans, that all set very specific boundaries delimiting the core and buffer areas of
the FRA, and determine specific regulations to deter illegal, unreported and unregulated (IUU) fishing.

---

jurisdictions, at global, regional, national, local/
community, cross-sectoral and sectoral levels.
The strategic aspects of governance often get
most attention at global, regional and national
levels, whereas the operational aspects become
increasingly prominent at national, subnational
and community scales.

The risk of failure of the OECM mainstreaming process in the fishery sector, in any
jurisdiction (EEZ, RFMO/A, TURF, LMMA), is
strongly influenced by the overarching governance system in such jurisdictions, with weak
governance increasing the risk of failure. The
governance processes that would affect OECMs
implementation and their biodiversity outcomes
depend on: (i) the political system in place with
its economic development and biodiversity conservation policies; (ii) the mechanisms in place
to facilitate collaboration between sectoral and
environmental institutions and among sectors;
(iii) the degree of decentralization and inclusiveness of decision-making power, in fisheries
and in conservation; (iv) the implementation
capacity available to the legitimate authority; (v)
the history of the relations among sectors and
between them and the conservation agencies
and interests; and (vi) the willingness to make
decisions for the greater public good rather than
more selfish reasons.

These factors are potentially relevant in
both central and local governance systems
but may be approached differently in each.[10]
For example, in Canada, if an Indigenous
government or community wishes to protect
an important ecological or cultural component
of biodiversity and seeks collaboration with
a federal department or agency to ensure its

protection through establishment of an OECM,
the proposed site would need to meet all the
criteria required in the Government of Canada
guidance, including the management of existing
and foreseeable activities, to ensure that risks to
biodiversity are effectively avoided or mitigated
(Government of Canada, 2022). In Colombia,
a core group composed of the Ministry of
Environment and Sustainable Development, an
NGO and a research institute has been established to oversee all OECM evaluations, involving external evaluators to enhance independent
evaluations.[11]

Although overarching legal and policy
frameworks may need updating to boost/
facilitate the OECM development, it is logical
to assume that for the fishery sector and for
each OECM in particular, the governance and
management details suggested in Decision 14/8
will need to fit within the existing local legal and
policy framework which also may have to evolve,
in a two-way co-evolution process. The degree of
flexibility and the case-by-case implementation
suggested by the Decision provide the decision-
making space needed for that co-evolution.

In Decision 14/8, governance is addressed
in many different places, implicitly in the definition of OECMs, the guiding principles and the
criteria, and explicitly in the Annexes. Annex I of
Decision 14/8 provides voluntary guidance on
the integration of OECMs into wider landscapes
and seascapes and their mainstreaming in economic sectors, *inter alia* to contribute to SDGs.
Annex II of the Decision provides voluntary
guidance on governance models for protected
areas, addressing issues related to: (i) legitimate
authorities; (ii) free and prior informed consent

(FPIC) and inclusiveness of Indigenous Peoples and local communities; (iii) diversity of State and non-State governance models; (iv) recognition of stakeholders' rights (including tenure rights) and responsibilities; (v) the broad range of ecological, economic, social and spiritual values to consider; (vi) effectiveness in achieving the expected long-term outcomes; (vii) equity in representation, procedures and distribution of costs and benefits; and (viii) flexibility for context-sensitive implementation.

Most of these issues relate to the 'good governance' principles that emerged in the mid-1990s in sustainable development strategies, at the United Nations level (see Graham *et al.*, 2003), and have gained momentum. They are not new to fisheries management and biodiversity conservation but their degree of implementation is highly variable.

Assuming there is an overarching decision at the highest level of governance to start a process of identification and use of OECMs in all relevant economic sectors, key actions taken at that high level, consistent with Decision Annex II, would facilitate the identification of OECMs and their coherent and consistent implementation in fisheries. Because of the context specificity and diversity of national jurisdictions, generalizations are inappropriate beyond stressing that many of the mainstreaming actions required in the fishery sector and described in the next sections would be facilitated if enabled by explicit but adaptive national policy frameworks giving fishery authorities the ability and incentives to engage in rolling out the process within the sector, in partnership with other sectors as appropriate.

The overarching activities that could facilitate OECM mainstreaming in the capture fisheries sectors include: (i) developing a vision or policy statement for the OECM initiative to support implementation at the appropriate governance level(s), in a highly participative process, to frame sectoral initiatives; (ii) reviewing and strengthening of the sectoral policy, legal and regulatory frameworks (gap analysis), including for empowerment of subnational governance systems and improvement of cross-sectoral co-ordination; (iii) mandating the 'legitimate authority' and clarifying responsibilities in OECMs, e.g. for decision making, identification, management, monitoring,

evaluation and reporting (MER), as well as mechanisms for cross-sectoral collaboration, conflict resolution and comprehensive reporting to the legitimate authorities and, as appropriate, to WCMC. Logically, the legitimate authority with a mandate to manage fisheries and take area-based management measures is the one that will be mandated to identify fishery-OECMs in its area of competence and report to the State, possibly in collaboration with the institution in charge of biodiversity. The reporting channel to CBD and WCMC is decided by the State and may often be the national focal point for biodiversity; (iv) developing or strengthening the collaborative processes among jurisdictions, economic sectors and at seascape level when relevant, in particular, co-ordinating the fishery-MER systems, including those covering large-scale and small-scale fisheries, and integrating them as needed with those of other sectoral management and biodiversity conservation agencies operating in the same area or in surrounding or functionally connected areas; (v) providing oversight and auditing to check the effective contribution of OECMs (see Section 7.7)

Additional complementary activities may be considered such as: (i) diffusing the generic guidelines on OECMs that are available (e.g. in IUCN WCPA, 2019) in national and local languages, adapting and translating them as needed for local use to raise awareness and interest; (ii) establishing a consistent national process for overseeing the handling of OECM proposals, across regions, sectors and types of governance. A competent, dedicated and highly visible team, mandated to help interested proponents in shepherding proposals through all the steps, could be effective; (iii) creating or updating a national database of all protected areas including MPAs, OECMs, LMMAs and other community-managed and sector-managed areas producing or likely to produce biodiversity benefits and co-benefits;[12] (iv) establishing accessible sources of funds and other implementation means, including for capacity building at local level (see more details in Section 3.1 and Table 3.1); (v) adopting a strategy and plan and a reasonable timetable for sectoral submissions of OECM proposals by the various sectors to the government or legitimate authority, if the OECM implementation process is deconcentrated at various levels; (vi) communicating on – and promoting – OECMs across the relevant economic sectors, as conservation mechanisms that are complementary to other conservation measures, compatible with sustainable use, and that bridge and help unify fishery and biodiversity conservation frameworks.

These actions require an implementation capacity which might be limited in many places, both centrally and locally, and co-operation may need to be developed or enhanced, at bilateral and regional levels. Following Decision 14/8, all these actions and all the actions referred to in the following sections of this guide are intended to be taken and implemented in multi-stakeholder processes involving all legitimate authorities, including Indigenous Peoples and local communities, with due consideration of their rights, responsibilities, institutions and values.

4.2 Fishery Governance Framework

Many of the actions considered below reflect, at the fishery sector and subsector levels, the actions listed above in the overarching national governance framework, which should incentivize and facilitate them. In cases where an ABFM is identified as a fishery-OECM, the legitimate authority who established the ABFM typically would have the sustainability of the target species and the fishery as primary objectives,[13] with due regard to habitats essential for the fishery and legally protected species. The area is identified as a fishery-OECM because it generates broader conservation benefits for the biodiversity of concern, including non-target species, threatened species, other critical habitats, the broader biodiversity, and ecosystem functions and services.

For mobile species, these benefits can spill over to the outside ecosystem, allowing the regulations in place within an OECM to improve sustainability and conservation performance of management not only within the OECM but also in the whole fishing ground and possibly ecosystem. However, for these species, the regulations in place and the activities allowed in waters around OECMs could also negatively impact the net benefit of the OECM for broader biodiversity if that threat is not explicitly addressed in the original FMP (see also sections 5.2.8 and

7.3.1). Consequently, appropriate regulations applied within and around OECMs will need to be complementary, coherent with both fisheries and conservation objectives, and integrated in the fisheries management plans. Such fishery-OECMs should also be integrated with the relevant biodiversity conservation networks and strategies, at local, national, regional or seascape levels, as required by species movements. The issue is addressed in Section 6.4.

In addition to the national governance framework highlighted above, States have the obligation to co-operate, including through regional fisheries management organizations or arrangements (RFMOs/As), for the conservation of living resources that straddle between EEZs and high seas, highly migratory species and high sea species (UNCLOS, Parts V, VII; UNFSA). RFMOs/As are the main means of co-operation for the conservation and sustainable use of these populations under UNCLOS and UNFSA. The ecosystem approach to fisheries is largely embraced under these two treaties through specific provisions although the term is not expressly incorporated.[14] These provisions are relevant to the identification of OECMs in a broader ecosystem approach context as previously discussed in this book. Furthermore, UNFSA contains several obligations for its Parties in fulfilling their obligation to co-operate through RFMOs/As with respect to their functions, including those related to agreeing and complying with their conservation and management measures, adopting and applying generally recommended international minimum standards for the responsible conduct of fishing operations, obtaining and evaluating scientific advice, and assessing the impact of fishing on non-target and associated or dependent species, among others (UNFSA, Art. 10). RFMOs/As operate in accordance with their mandates reflected in their respective conventions. Several RFMOs have updated their conventions through amendments in light of the UNFSA provisions. The examples of RFMOs' OECMs provided above (i.e. NAFO and NEAFC) are those well equipped to identify OECMs against the CBD Criteria in a robust manner if they choose to do so and based on sound scientific assessment and advice, in both cases even though other impacts on the potential OECM area from activities other than fishing were also considered to adhere with the

CBD Criterion B that states that 'the Governance may be by a single authority and/or organization or through collaboration among relevant authorities and provides the ability to address threats collectively' (CBD Decision 14/8, Annex III, Criterion B).

The following actions expected from the fishery sector would allow a review and strengthening of the fishery policy, regulatory frameworks as needed, at the appropriate governance levels (the legal framework is examined in Section 4.3). The actions are presented below in a logical order, but the need for such actions, and the order in which they may be taken, strongly depends on the strengths and weaknesses of the existing fishery governance and management systems, justifying the implementation 'flexibility' recommended in Decision 14/8.

4.2.1   Mainstreaming OECMs in the fishery sector

(i) Review and revise, as necessary, the existing legal, policy and budgetary frameworks of the sector to facilitate mainstreaming (Garcia *et al.*, 2019, 2020a; Marnewick *et al.*, 2020: 14). (ii) Encourage the fishery sector, providing States with economic and other incentives to identify new opportunities and recognize the contribution of existing OECMs to sustain or improve ecosystem functions and services. (iii) Facilitate building up of capacities required to improve mainstreaming of OECMs, including in assessment, management, monitoring and evaluation, under the diverse modes of governance of the sector.

4.2.2   Establishing or identifying an auditing authority and process

Independent auditing of management performance is not yet as widespread a practice for most national fishery management regimes as it is for fisheries certified under the Marine Stewardship Council (MSC)[15] and most RFMOs/As.[16] However, systematic and regular performance evaluation improves performance of adaptive fishery management systems, and formal auditing (whether internal or third party) would add

credibility to reports on OECMs' performance. The use of OECMs is therefore an opportunity and incentive to establish or strengthen performance assessment for the fisheries and the sector and not only for OECMs (see Section 7.3).

4.2.3   Operationalizing equitable governance

In line with the well-accepted principles of 'good governance' (see Section 5.5.2.2), Decision 14/8 recognizes the need for equity at three levels: (i) recognition of the legitimate authority and all legitimate actors, with their gender, identity, rights, values, knowledge systems and institutions; (ii) procedures, giving effect to the 'recognition' by ensuring inclusive institutions and mechanisms from data collection to decision making and implementation; and (iii) distribution of costs and benefits (of the OECM and the fisheries) among actors.[17] The Ecosystem Approach to Fisheries (FAO, 2003) and the FAO Voluntary Guidelines for Securing Sustainable Small-Scale Fisheries in the Context of Food Security and Poverty Eradication (FAO, 2015, Chapter 5B) are good sources of guidance in that respect, stressing the link between the right to access resources and the responsibility to manage them and conserve biodiversity.[18] It can be challenging to ensure equity in participation and distribution of opportunities, costs and benefits among actors in relation to sustainable use and conservation of biodiversity. These different actors may value biodiversity attributes and benefits differently. Some resource harvesters may feel no responsibility to share the benefits from fishing with biodiversity conservation interests, and some biodiversity conservationists may feel no responsibility to share the costs of preventing damage to biodiversity features.

4.2.4   Facilitating co-ordination/ integration to improve performance

Consider the needs, cost and benefits of integration of sustainable use and conservation of biodiversity into the vision, goals and targets of fisheries and conservation policies and regulations at national and regional scale.[19] Parties to the 1995 UNFSA[20] already accepted the obligation to protect marine biodiversity with UNFSA Article 5 (g). This provision is applicable in areas within and beyond national jurisdiction as per Article 3 of UNFSA.

Integration of OECMs and their management must be done at multiple governance scales: (i) within the fishery in which they are established, integrating them into its fishery management plan (FMP); (ii) within the fishery sector, across the different fisheries using a given ecosystem; (iii) among economic sectors potentially impacting the same OECM (e.g. fisheries, navigation, oil and gas, mining, tourism and renewable energy), e.g. through marine spatial planning; (iv) in national poverty eradication and sustainable development strategies (SDGs, etc.), in relation to ecosystem services (including at least provisioning and cultural services); (v) across jurisdictional boundaries (e.g. for shared or straddling OECMs); and (vi) at ecosystem or ecoregional level, in existing seascapes, EBSAs,[21] etc. Levels of integration (i) and (ii) are achievable within the fishery sector. Levels (iii) and (iv) require action by the State. Level (v) requires international collaboration, and level (vi) may be possible at subnational, national or international level, depending on the nature and size of the ecosystem.

4.2.5 Identifying negative impacts of fisheries on OECM outcomes

When considering OECMs and other marine conservation efforts, it is important to assess the non-sustainable impacts by fishing, not only in conservation areas but in the entire fishery sector. This is consistent with other fishery commitments in: (i) the EAF; (ii) avoidance of significant adverse impacts on vulnerable ecosystems (UNGA Resolution 61/105; FAO, 2008, 2009b); and (iii) avoidance of any destructive fishing practices (FAO and UNEP, 2010; McCarthy et al., 2024). Destructive fishing on all scales is already prohibited under the 1995 FAO Code of Conduct for Responsible Fisheries the 2004 UNGA Resolution 59/25 on sustainable fisheries (Article 66), the 2012 Rio + 20 UNCSD Summit (§168 of The Future We Want), the 2015 FAO voluntary guidelines on small-scale

fisheries (Art. 5.16) and the 2016 Sustainable Development Goals (Target 14.4) (Willer et al., 2022). In seeking to identify and limit negative impacts of fisheries, Decision 14/8 lists the requirements for OECMs and encourages sectors (including fisheries) to identify OECMs in their areas of competence. It does not specify the *type* of fishery suited to a fishery-OECM.

It has been considered that 'industrial' fisheries, and indeed any industrial activities, should be excluded from MPAs and OECMs as not compatible with their conservation objectives (Day et al., 2012, 2019; IUCN-WCPA, 2019: 6).[22] It is already internationally agreed that any fishery, at any scale large or small, must be regulated with enhanced risk aversion for SAIs in general and particularly in vulnerable areas. However, there is no general agreement on 'industrial fisheries' definition or vessel size, and Decision 14/8 does not make any reference to the scale of the economic sectors in which it calls for OECMs when it suggests to 'identify and prioritize the sectors most responsible for habitat fragmentation, including ... fisheries ... to engage them in developing strategies for mitigating the impacts on protected areas and protected area networks including OECMs...' (Annex 1, p. 4e). Moreover, it is logical to establish effective OECMs in all intensively fished areas, irrespective of fisheries' scale, to mitigate their collateral impact, and restriction of their access to conserved areas currently remains a decision of the State or legitimate authority, based on objective local information and the OECM identification criteria (see Appendix 4 for additional considerations).

4.2.6 Identifying negative impacts of other sectors on fisheries and OECM biodiversity conservation outcomes in these fisheries

Seabed mining, oil and gas industries, land-based pollution and navigation are examples of such threats. Such issues may be addressed within cross-sectoral frameworks at an appropriate spatial scale, like integrated coastal zone management (ICZM), marine spatial planning (MSP) or equivalents. These can aim to harmonize OECMs with other relevant spatial and non-spatial measures applied by other sectors or biodiversity conservation agencies in the same area, to increase synergy and reduce conflict. Where such cross-sectoral initiatives are not ongoing, the fishery sector could take the initiative towards improved sector-based conservation and be an example that other sectors might follow. In many instances, a bilateral collaboration between two sectors may be enough to make an OECM operational and even to establish cross-sectoral OECM outcomes, as long as the State or the RFMO recognizes (and, ideally, supports) such collaboration. In the case of NAFO, one of its VME closures, that would in principle qualify as OECM, has not been identified as such by NAFO contracting parties (at least initially) because of the potential impacts caused by other human activities, such as oil and gas exploration, in the same area (NAFO-COM-SC, 2023).

4.2.7   Ensuring that an effective fishery management system is in place

To be effective, a fisheries management system needs to be appropriate for the specific setting in which it operates. States usually try to manage all their fisheries effectively but because of the range of ecological, technological and socio-economic realities, their management can vary greatly in capacity, sophistication and effectiveness. What constitutes an effective management system for medium- to large-scale fisheries may involve fishery-specific access rules, gear regulations, area-based measures, and effort and catch limits. The small-scale fishery sector, with its complex set of fishing targets, gears and strategies, is a multispecies and multigear reality (particularly important in developing nations), involving many forms of management, e.g. centralized in the authority headquarters, decentralized in regions or municipalities, or devolved to Indigenous Peoples and local communities, usually under various forms of co-management. The capacity of the fishery to effectively use and manage a fishery-OECM must therefore be assessed.

The explicit integration of expected biodiversity conservation outcomes of the OECMs into the fishery management plan (FMP) of a large fishery or the fishing practices of small-scale fisheries (SSFs) is a significant move in mainstreaming biodiversity concerns in the sector. It can be noted in that respect that while there may be biodiversity-related additions required in a management plan for a large-scale fishery, such aspects may already be present in SSFs managed by local communities and using traditional/cultural practices (IPBES, 2022a).

4.2.8   Adopting or strengthening the EAF as the operational framework for managing fisheries and OECMs

The EAF has already been adopted by the FAO (since 2001) and all advanced countries and RFMOs/As, recognizing the need to consider and limit the impact on non-target resources and habitats. As the features and operations of EAF continue to improve and be refined, the identification and inclusion of fishery-OECMs in EAF-based management plans should facilitate their expanded use in fisheries. Making OECMs an explicit part of the EAF would in turn strengthen EAF implementation, giving more specificity and priority to biodiversity outcomes, and more long-term security to EAF measures. The FAO guidelines on EAF (FAO, 2003, 2009) could be amended or supplemented to explicitly address the conservation aspects of OECMs.

4.2.9   Strengthening the monitoring and evaluation capacity

There are multiple approaches for OECM monitoring and evaluation. On the one hand, in local-level or community-based OECMs, such processes may be developed within a decentralized or co-management framework. On the other hand, for government-managed OECMs, agencies and processes that have scientific capacity for stocks assessment and management advice (e.g. national fishery research laboratories) can be tasked and enhanced if needed to identify and monitor OECMs and assess their performance. The establishment or strengthening of collaborative monitoring, research and assessment programmes between fisheries management and biodiversity conservation agencies, at national and regional levels, would be an enabling factor. This could also engage with and draw upon local and Indigenous knowledge, including appropriate monitoring processes. Co-ordinating the expertise and programmes for assessing and managing MPA networks and related biodiversity initiatives with those monitoring and evaluating fisheries performance (MER; see Chapter 7), in the same area or ecosystem, could both optimize overall workload and help mobilize additional resources.

4.2.10   Identifying the need for transboundary collaboration for OECMs located across different jurisdictions

Using the stocks-based terminology, fishery-OECMs might be 'shared' (the features of the OECM extending into neighbouring EEZs), 'straddling' (overlapping one or more EEZs and the high sea) or implemented entirely in specific sites in the high seas. Effectiveness would benefit from international collaborative action for assessment and management. Within their EEZs, formal shared stocks agreements are uncommon but have been negotiated among EU Member States, between Norway and Russia, and the USA and Canada. On the high seas, there are a few cases of collaborations on biodiversity issues between RFMOs/As and regional seas conventions (RSCs), like the collaboration between NEAFC and OSPAR or between GFCM and the Barcelona Convention. The several forms of transboundary OECMs may promote – and possibly require – improvement in such co-operative arrangements.

4.2.11   Matching implementation capacity and commitments

The institutional, scientific and management capacity building required to deal with OECMs within fisheries may call for additional resources for science and management in a sector that often is chronically underfunded. In many cases, ambitions will need to be tailored to the resources available, creating an incentive for user sectors and conservation agencies to pool knowledge and resources to inform more robust OECM assessment and management either (i) assessing all potential OECMs at once or (ii) using

pilot experiments to learn by doing, making full use of the knowledge of Indigenous Peoples and local communities. However, the weaker the evidence available, the less likely it is that a positive decision on OECM status can be made. Furthermore, the weaker the management efforts, the less likely it is that expectations of long-term benefits can be justified. In countries or fishing areas under weak governance and chronic overfishing, ABFMs are unlikely to meet the OECM criteria. The issue of capacity building is examined in more detail in Section 3.1 and Table 3.1.

The list of actions above may look intimidating, but every point mentioned would be needed anyway for any substantial improvement in conventional fisheries management, as these actions would improve the whole management system and not just the fishery-OECM performance.

4.3   Fisheries Legal Framework

The international framework enabling the identification and implementation of OECMs in all ecosystems has been established by Decision 14/8 which encourages governments to '… identify other effective area-based conservation measures and their diverse options within their jurisdiction' (§5a). It also recognizes the '… potentially different legal regimes for different portions of the same marine areas (e.g. seabed and water column in marine areas beyond national jurisdiction)' (Annex IV, A, 1, j). According to the Convention, CBD provisions and COP Decisions apply only to CBD Parties, within areas under their jurisdiction, when referring to biodiversity components such as species or genetic resources (Art. 4a) for which they have responsibility, and within and beyond the area of its national jurisdiction when referring to biodiversity-related processes and activities (Art. 4b). In addition, under UNCLOS (1982, Art. 63.1 and 64.1) and the UN Fish Stock Agreement (1995 Art. 1b and 5b, c, g), States members of RFMOs/As co-operate and agree on measures for conservation, management and sustainable use, in and beyond their EEZ. Such measures presently include ABFM measures of various types, some of which protect VMEs.

Nothing in international law or in the CBD impedes CBD Parties from proposing to RFMOs, to which they are also Parties, to adopt fishery-OECMs in line with the CBD guidance and this is already ongoing in NAFO, NEAFC and GFCM (as discussed above and illustrated in Boxes 2.1, 4.1 and 4.2). After coming into force, the BBNJ Agreement will certainly influence the way in which and extent to which ABMTs, including MPAs and OECMs, may be used in the future by any sector in the ABNJ, possibly fostering cross-sectoral collaborations.

To become fully operational in fisheries, the actions explicitly expected or implied in Decision 14/8 would need to be consistent or reconciled with the existing fisheries regulatory framework, at national level in the fisheries legislation, at subnational level, e.g. in the traditional management systems of Indigenous Peoples and other local communities, and at regional level, in RFMOs/As. Under existing frameworks, area-based management measures have been used for centuries (e.g. Hindson et al., 2005; Die, 2009; Cochrane and Garcia, 2009; NOAA, 2017; Rice et al., 2018; Garcia et al., 2023). Their purpose sometimes has been narrowly to contribute to ensuring long-term conventional sustainability of the fishery and its target species. However, consistent with the ecosystem approach to fisheries, they have also been established to reduce or avoid by-catches of unwanted or protected species, and to protect sensitive[23] and essential[24] habitats.

ABFMs found or upgraded to meet Decision 14/8 criteria as OECMs may be implemented in fisheries with little modification, if any, of their present regulatory frames. However, complementary actions to further solidify the intended outcomes of OECMs may include the following.

- *Mandating the legitimate fisheries management authorities.*[25] For marine capture fisheries, the mandated authorities with the right to adopt and enforce management measures (including area-based measures) are usually already defined in countries with some sort of Fisheries Act, either at central State level (ministry, department) or other levels (federal States, Indigenous People, local communities, associations, etc.). In the high sea, flag States, individually

or collectively through RFMOs/As, jointly have the required competence.

- *Including OECMs as management instruments in fisheries legislation* may help to reach consistency across fisheries with related concepts such as 'sustained management' and 'long-term biodiversity outcomes', and signalling the formal intent to use OECMs and produce their expected biodiversity outcomes in the long term. Although the concept of 'long term' was not bounded in Decision 14/8, the legitimate authority of fishery-OECMs must have a clear commitment to maintain the biodiversity outcomes for as long a time as necessary. This implies maintaining the ability to adapt OECMs characteristics to important contextual environmental or socioeconomic changes. Such a commitment is necessary because the long-term intent of a fishery and its management to maintain the ABFM (with its technical regulations) to sustain the harvests does not automatically guarantee the 'long-term' success of the regulations in delivering the biodiversity outcomes expected from the OECM, in the same area. Moreover, if any new provision of the fishery management strategy or plan would appear to threaten the long-term nature of the biodiversity outcomes of an otherwise effective fishery-OECM, the OECM status should be reassessed and appropriate action taken. Simple, universal standards of management 'sustenance' cannot be defined but evidence likely to be necessary to ensure the 'long term' includes (i) the designation of a legitimate management authority with a long-term mandate; (ii) a formal policy or legal provision stating the 'long-term' intention, clarifying the process and conditions needed to change the OECM area or regulations and estimating the likelihood that such changes might happen; (iii) setting explicitly long-term objectives documented in an official publication from the legitimate authority; (iv) formal adoption of a management plan or traditional equivalent form, with appropriate provisions, in each fishery; (v) showing that necessary financial and human resources are adequate and secure in the long term; (vi) establishment of a MER system demonstrating the long-term monitoring and assessment capacities; (vii) co-ordination of the OECM management with other conservation efforts of agencies with authority for biodiversity conservation. As an example, in Canada (CCFAM, 2017:20, §4):

the measures identified as OEABCM (herein referred to as OECM) will be managed using a long-term adaptive management approach

and are expected to be in place year-round for a minimum of 25 years to support long-term biodiversity conservation benefits. This criterion should not be considered an expiry date for OEABCM. The underlying aim is for all reported OEABCM to be in place indefinitely and ideally in perpetuity.

- *Elaborating additional regulation on OECMs that could protect fishery-OECMs from negative impacts on biodiversity from other human activities, or establish rules for elaborating cross-sectoral OECMs.*

Notes

[1] Both the UNFSA and the BBNJ Agreement are implementation agreements of UNCLOS.

[2] The 'high seas' are the ocean waters areas beyond national jurisdiction while the 'area' is the seabed and ocean floor and subsoil thereof, beyond national jurisdiction.

[3] www.neafc.org/oecm (accessed 21 July 2025).

[4] Lyell Centre, Heriot-Watt University, Edinburgh, Scotland.

[5] Executive Secretary, North-East Atlantic Fisheries Commission (NEAFC), London, UK.

[6] www.cbd.int/soi (accessed 21 July 2025).

[7] www.ospar.org/about/international-cooperation/collective-arrangement (accessed 21 July 2025).

[8] GFCM Secretariat.

[9] Like in the case of NEAFC OECMs (ICES, 2023)

[10] For LMMAs, see for example Govan et al. (2008).

[11] https://panorama.solutions/en/solution/beyond-protected-areas-recognition-oecms-colombia

[12] Following the Decision (p. 14) 'benefits' are *intended* (hence related to explicit objectives) while 'co-benefits' are *unintended*, i.e. obtained incidentally, while pursuing another objective or simply not considered as objectives.

[13] Some ABFMs are used to regulate access, allocating space to subsectors, to allocate resources or reduce sources of conflicts and accidents, e.g. separating artisanal from industrial fisheries or set gears from mobile ones. In VMEs, the biodiversity conservation objective is particularly prominent, but the link between that protection and the productivity of the target species is present.

[14] For instance, EAF principles can be identified from a range of UNCLOS and UNFSA provisions including UNCLOS Arts. Arts 61(2), (3), (4), 63, 64, 119(1) (a) and UNFSA Arts 5(a), (b), (e), (h), (j), 6 on avoiding overfishing and rebuilding populations; UNCLOS Arts 192, 194(5) and 204-206, and UNFSA Arts 5(d), (f) and 6, with respect to minimizing fishing impacts; UNCLOS Arts 61(3), (4), 63, 64, 119(1)(b) and UNFSA Arts 5(b), (e), (j), 6, 7 with respect to species interactions considerations and compatibility of measures within and beyond national jurisdiction; UNCLOS Arts 61, 63, 64, 119, 192, 193 and 194 with respect to maintaining ecosystem integrity, and UNFSA Arts 6 and Annex II, 5(g), 10, in relation to the application of the precautionary approach and maintenance of ecosystem integrity.

[15] The Marine Stewardship Council (MSC) and the Aquaculture Stewardship Council (ASC) have partnered to share the Chain of Custody Standard. This means one certification audit covers both MSC and ASC products.

[16] Most RFMOs now undertake overall performance assessments using either external reviewers or third-party auditing.

[17] Noting that 'equity' has to take into account potential differences among actors in ability to pay for costs and vulnerabilities to decreases in benefits.

[18] This is also consistent with CBD Decisions V/6 (2000) and VII/11 (2004) on the ecosystem approach and respective guidance for implementation.

[19] It is important to note that Parties to the 1995 United Nations Agreement for the Implementation of the Provisions of the United Nations Convention on the Law of the Sea of 10 December 1982 relating to the Conservation and Management of Straddling Fish Stocks and Highly Migratory Fish Stocks (UN Fish Stocks Agreement or UNFSA) have the obligation to protect marine biodiversity in areas within and beyond national jurisdiction (Articles 3, 5g).

[20] United Nations Agreement for the Implementation of the Provisions of the United Nations Convention on the Law of the Sea of 10 December 1982 relating to the Conservation and Management of Straddling Fish Stocks and Highly Migratory Fish Stocks (UN Fish Stocks Agreement or UNFSA).

[21] Ecologically or Biologically Significant Marine Areas (EBSAs).

[22] Between the 2012 and 2019 versions of these MPA guidelines, the reference to industrial activities has increased eight times (from 3 to 25). For a discussion of the concept of 'industrial' fisheries, see Appendix 4.

[23] Sensitive habitats are habitats that are vulnerable to fishing activities and important for ecosystem functions and services.

[24] Essential habitats are habitats that are needed to maintain the productivity of the fishery target species.

[25] For co-ordination with the CBD, the legitimate authority is usually already established as a focal point in the ministry in charge of biodiversity.

5   Identification of Fishery-OECMS

Abstract

In this central chapter, the initial identification phase of the fishery-OECM implementation cycle is described in five interconnected quasi-sequential steps: (i) consolidation of the information, (ii) quick screening of ABFMs, (iii) full assessment of potential fishery-OECMs, (iv) synthesis and reporting to legitimate authorities, and (v) decision making about OECM status. After reviewing the premises of this phase, the different steps are described in detail, providing clarifications regarding the common principles, criteria, subcriteria and suggested 'elements of evidence' needed for the assessment, when considered in capture fisheries. The chapter highlights the need for a thorough, robust and transparent assessment as well as the specific difficulties emerging in the ocean and in the sector.

The identification phase is the first phase of the fishery-OECM implementation cycle briefly described in Chapter 3. The enabling legal and policy frameworks referred to in Chapter 4 ought to be in place before the identification phase, to facilitate the process. In some countries, the strong fisheries frameworks in place are a good starting point, and with so many ABFMs already in place, the identification of fishery-OECMs may start at any time. Even if some fine-tuning were needed by the legal and policy frameworks, there could be sufficient information on biodiversity outcomes and effectiveness to allow the legitimate authority to conduct evidence-based evaluations of fishery-OECMs using the OECM criteria. In Canada, for example, the formal identification of marine OECMs began even before the CBD COP approved Decision 14/8 (CCFAM, 2017; Hiltz *et al.*, 2018; Aften and Fuller, 2019). Since 2018, the identification of OECMs in the coastal and marine fisheries domains has been progressing (Jorgensen *et al.*, 2020; ICES, 2021; Shackell *et al.*, 2021). In community-managed areas, informal traditional frameworks should already be in place to start a participative identification process, with subsequent engagement by the State and possibly NGOs involved in conservation, development and capacity building, and other interested parties, depending on the governance customs and policies of the region.

The different steps of the process are detailed below.

5.1   Reflections on the Identification Phase

The identification phase of the fishery OECM implementation cycle (given in Fig. 3.1) is intended to assemble, review and prepare the advice that the legitimate authority needs in order to decide which of the existing or planned ABFMs could be referred to as 'fishery-OECMs' with all the potential pros and cons of such a decision.

A detailed representation of the phase is given in Fig. 5.1, in five sequential steps:

1. consolidation of the information
2. quick screening
3. full assessment
4. synthesis and reporting
5. decision.

The decisions are subsequently transmitted to the management authority for integration of fishery-OECMs in the relevant fisheries management plans. The decisions may also be reported to the WCMC database, if decided by the State/authority. It is important to stress, however, that

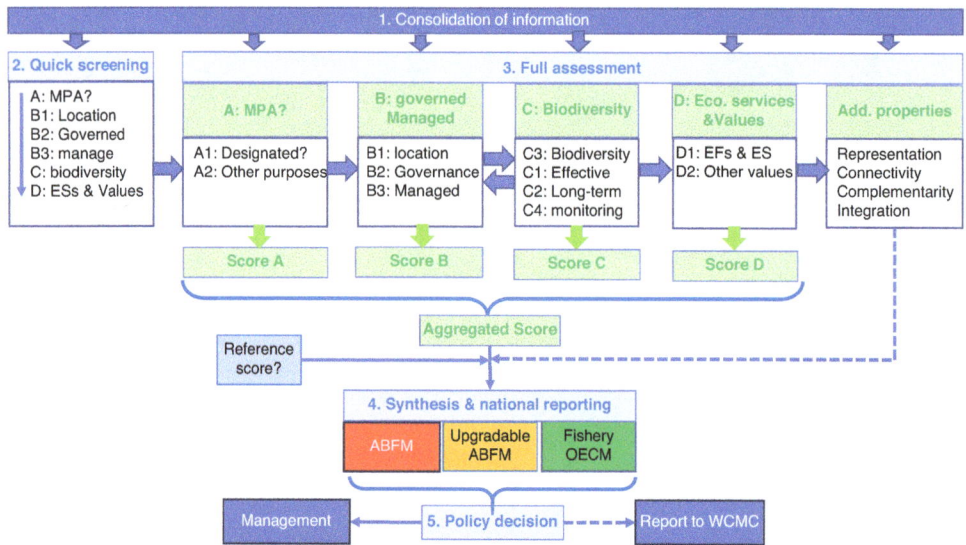

Fig. 5.1. Graphical representation of the stepwise identification process. The order of the subcriteria C1–C4 has been modified to fit the order in which the elements of information are needed. The two-way arrows between criteria B and C indicate that assessors may have to move back and forth between the two when progressing in the full assessment.

any fishery-OECM not officially reported will not be accounted for in the international targets on conservation coverage (like the GBF Target 3).

The list of criteria, subcriteria and suggested elements of evidence provided by Decision 14/8 (Annex III) is presented in Section 2.4 and Appendix 2 and is central to conduct of the phase.

In the identification phase, ABFMs passing the quick screening test will be considered potential OECMs and submitted to a full assessment, criterion by criterion. Those satisfactorily meeting the criteria will be considered candidate OECMs and presented to the legitimate authority for decision and formal identification as fishery-OECMs. Those ABFMs which satisfactorily meet many but not all requirements might be considered as upgradable ABFMs. These may be presented to the legitimate authority for decision regarding their possible upgrading, considering biodiversity attributes in the area, feasibility and costs of the regulations and changes needed to improve performance against the criteria, and benefits to biodiversity, the fishery, and the authority from achieving OEFM status for the area. In addition, new ABFMs may be established with the intent of meeting the OECM criteria,

with the positive biodiversity outcomes as a priority objective, following the same identification process. At the end of the identification phase, fishery-OECMs will have been formally identified by the legitimate authority. If decided, the regulations justifying OECM status will be integrated in the fisheries management plan, if not already part of it, and implemented within it.

The detailed identification steps suggested in Fig. 5.1 and below follow the order of the criteria given in Decision 14/8 (Annex III) and reproduced in Appendix 2. However, there is no requirement in Decision 14/8 to follow that particular order as long as all criteria are considered. The list of criteria may be seen as a checklist of properties essential to OECMs that must be evaluated by the end of the process, and when synthetizing the conclusions of the assessment to be considered by the legitimate authority for final decision.

In applying the criteria, the following steps are necessary.

1. Ensure that the area is not an MPA (criterion A) because otherwise it cannot be an OECM and there is no point in proceeding.

2. Get agreement on the geographic area to consider for OECM status (subcriterion B1) to bound subsequent discussion of the biodiversity and management features of the area, and to be able to calculate the surface of the area, to be accounted for in the global coverage target.

3. After these initial steps, in waters with well-established fisheries management plans and agencies (satisfying subcriteria B2 and B3), it may be appropriate to next determine what significant positive biodiversity outcomes are occurring or expected (subcriterion C3), because the nature of those outcomes may inform what factors are crucial for 'effectiveness' (subcriterion C1) and which actors must be 'equitably' engaged in governance (subcriterion B2). However, there are also marine areas likely to be rich in biodiversity (subcriterion C3) but where management capacity has been historically weak or compliance with regulations is unknown or unreliable (subcriterion B3). In such areas, establishing how equitable and effective governance and management are may be more crucial for the OECM evaluation than agreeing on the relative priorities among potential positive biodiversity outcomes of effective management and governance.

4. In all cases, evaluating the attention given to ecosystem services and functions, local knowledge and other locally relevant values (criterion D) may follow the evaluation of all the other criteria, although the evaluation of criterion B often inherently includes discussion of ecosystem services and functions, and evaluation of criteria B and C often may be where local knowledge and values do or don't receive appropriate attention.

Given this diversity of possible circumstances that may be encountered, each assessment team may, after consideration of the guidance available and the local conditions, develop its own identification pathway through the criteria. For each criterion, it will be necessary to figure out if enough unequivocal information is available to decide whether the criterion is met. This process is facilitated by the fact that the criteria given in Decision 14/8 are subdivided into more homogeneous subcriteria, each accompanied by a short list of elements of information (or qualitative indicators) that can guide the evaluation (see Appendix 2).

The evaluation of the degree to which each criterion is met will rarely be a simple yes or no, but in most cases a level of performance within a range from poor to excellent. Decision 14/8 gives no guidance in how performance on the various criteria should be combined to guide a final decision. Possible approaches are discussed in the synthesis of the assessment conclusions (Section 5.6).

The only important requirement in all cases is to have examined all criteria at the end of the process, using the best available information to give the best possible assessment to the legitimate authority about the OECM candidate. In cases where the evaluation concludes that all criteria are sufficiently met and OECM status would be appropriate for the area, the assessment report should include all important evidence and analyses, with technical details in annexes or otherwise, readily accessible to users.

In cases where the evaluation concludes that one or more criterion is not met, or could only be met with upgrading, evaluation of other criteria may or may not have been exhaustive. Nevertheless, all relevant information and considerations taken into account in the assessment should be summarized in the report and fully archived, so they can be available should the legitimate authority pursue the upgrading and then want another assessment for OECM status. This may be particularly the case when a group of experts in a field were assembled for an assessment, and their interactions during the workshop and assessment increased their common understanding of the biodiversity, management and/or governance of the area, even if it does not meet all OECM criteria.

It is also very helpful to write the report of the assessment following exactly the list of criteria in the order given in Decision 14/8. Such a standard report structure facilitates coherence between different groups of assessors (e.g. in inland and marine domains), and when evaluating progress towards overall targets, reviewing and learning from experiences with the various criteria and with different types of information, and in co-ordinating management and conservation efforts across multiple OECMs and adjacent protected and 'open' areas.

Evidence		Sub-criteria		Criteria		Area
A	3	A	3	A	3	
B1a	3	B1	3			
B1b	3					
B2a	3	B2	2.5			
B2b	2					
B2c	2			B	2.7	
B2d	3					
B3a	2	B3	2.5			
B3b	3					
B3c	2					
B3d	3					
C1a	3	C1	2.8			2.9
C1b	3					
C1c	2					
C1d	3					
C2	3	C2	3	C	2.9	
C3	3	C3	3			
C4a	3	C4	2.8			
C4b	2					
C4c	3					
C4d	3					
D1a	2					

Evidence	Colour Code	Score	Range	Decision
Good		3	2.6-3	OECM
Medium		2	1.6-2.5	To upgrade
Poor Absent		1	1-1.5	To drop

Fig. 5.2. Some relationships among subcriteria (left) and elements of evidence (right). The overlaps and partial duplications between elements of evidence 'belonging' to different criteria, particularly between criteria B and C, indicate that: (i) these criteria are not independent; (ii) the list of elements of evidence is an important checklist but not a sequential assessment roadmap; and (iii) efforts to streamline the identification process may lead to improved efficiency in the use of the information and sets of competence.

The potential interactions among the criteria (in terms of the elements of evidence they call for) (Fig. 5.2) argue for using one comprehensive multidisciplinary team for the whole assessment of an OECM, and also that it would be useful to assess the criteria, particularly criteria (B) and (C), in parallel, if not jointly, sharing the information and multidisciplinary competence available in quasi real time. As an illustration of the connection between regulations and their expected outcomes, in the ICES-FEG meeting (ICES, 2021), the experts argued that, in some cases, clearly effective enforcement of the regulations (e.g. a ban on bottom-contacting gear) could be a good enough indicator that the expected positive biodiversity outcome (bottom habitat protection) is highly likely to be produced.

The stepwise process followed below (sections 5.3–5.7) contains steps additional to the list of criteria, that are needed to prepare and complete the criteria-based process. Such steps occur: (i) before addressing the criteria, to prepare the information and undertake a preliminary quick screening; and (ii) after having gone through all the criteria, to assess 'additional properties' of OECMs (not explicitly referred to in the criteria but stressed in the principles or voluntary guidance), and also to address the issue of synthesis and reporting of the assessment and underline the importance of the final decisions by the legitimate authority.

The multiple branching of many of the steps is consistent with the flexibility of the assessment pathways in different circumstances as described above, stressing that all pathways must consider all the criteria and subcriteria.

5.2 Premises

The suggested identification process described below is guided by some premises that stem mainly from the guiding principles contained in Decision 14/8 (Annex III, A), but also from experiences accumulating from the various meetings on fishery-OECMs referred to earlier. These premises must be kept in mind when proceeding through the different implementation phases and are therefore examined first below.

The identification process requires a good understanding of the concepts and terminology used in Decision 14/8 and in this book. Misunderstandings may arise for several reasons, such as when the Decision gives very specific meaning to terms that may be used widely in various colloquial or technical ways in research and management. Even terms not constrained by the wording of the Decision nevertheless may be used by various disciplinary specializations with meanings that may overlap but not be identically bounded. Experience showed that when organizing workshops of experts on OECM identification and use, terminology was often a problem. It is therefore strongly advised to familiarize all participants with the terminology at the start of a meeting, particularly terms with specific meanings in the Decision or known to often be interpreted differently in research and management, or between experts in conservation biology and in sustainable use of resources.

The identification process aims to examine the extent to which the potential OECM meets the definition and the guidance contained in the Decision. The core text of Decision 14/8 (pages 1 and 2) qualifies the guidance contained in the Annexes of the Decision as 'voluntary' (§1) to be applied in a *flexible way* and *case by case* (cf. §3). These terms imply that (i) the decision on whether and how to apply the voluntary guidance rests with the State Party to the CBD; and (ii) the national guidance may be adapted by legitimate authorities to their specific circumstances

and these of the sectors and areas concerned. In the process, the authorities are expected to remain faithful to the intent of the Decision and *consistent with the convention* (§4). This 'flexibility' is particularly important considering the large range of implementation capacity available in the world fisheries. However, no matter what resources are available, all criteria need to be considered, even though case by case the quality of the 'best available' data and methods can be much higher (or lower) for some criteria and subcriteria than for others.

The following considerations should receive attention before starting the identification process to facilitate and guide the assessment.

5.2.1 Comprehensive versus incremental approaches

Given that Decision 14/8 has practical, scientific, economic and political implications, the identification process may require significant resources. Consideration is needed as to whether it should be undertaken using a 'comprehensive' or an 'incremental' approach.

A comprehensive approach would be developed across the entire sector (and hence all the waters under national or RFMO/A jurisdiction, or major portions of them), cataloguing all existing ABFMs in the EEZ or in a large region. Each area entered in the catalogue would be 'scanned' for significant potential biodiversity benefits, listing all of them, and when the list is complete, assessing each ABFM against the criteria for identification. This systematic approach may generate economies of scale and improve coherence within and across fisheries and at ecosystem level. However, the necessary data and information mobilization would be extremely resource-demanding for legitimate authorities responsible for large ocean or coastal areas, and require a very large number of experts, including holders of local knowledge, at the workshops, to ensure there was 'best available knowledge' for each area. There could be substantial value in initial rapid overviews of information on biodiversity and management in the larger area, to gain early buy-in by the sector or environmental critics.

An incremental approach might be appropriate, selecting one or a few fisheries at a time, starting with ones with high sustainability and biodiversity conservation stakes, that are using (or could use) ABFMs reasonably likely to meet the standards for OECMs, and for which necessary information and expertise are more readily assembled. Starting with areas with well-studied biodiversity and well-documented management would allow a learning-by-doing process, progressively developing the capacity needed to fully mainstream fishery-OECMs in the sector. The process to upscale to the whole sector might be slower but it might be more feasible in capacity-limited fishing nations, or when legitimate authorities deal with many fisheries and target species. Under any approaches, the participation of fishers' organizations and committees is essential for buy-in, and participation by environmental NGOs (ENGOs) and/or other biodiversity experts is essential for subsequent credibility of any decisions.

5.2.2 Case-by-case evaluation

In line with Decision 14/8 (§3), the Introduction of Annex III, and Annex IV, D, d, and irrespective of the approach selected (comprehensive or incremental), ABFMs should be evaluated individually, case by case, mainly against the criteria for identification but considering also the guiding principles and voluntary guidance contained in the Decision. The reason for this is that the effectiveness of the different types of area-based measures at delivering biodiversity conservation outcomes – and hence the likelihood that they will meet OECM standards – depend on too many factors to be simply attributed *a priori* to categories of measures or types of areas (Rice and Garcia, 2018). The same consideration justifies not generalizing the results obtained in any single context to other contexts.

A case-by-case approach for each potential OECM is recommended to check, for example, application of the ecosystem and precautionary approaches; integration of the OECM with other area-based measures; use of the best information available; the range of positive biodiversity outcomes expected from the ABFM, the specific threats in the area and outside it when they

can impact the likelihood of the expected positive biodiversity outcome; and the governance processes (including assessment and decision making).

Consideration should also be given to the fact that regulations intended to generate positive outcomes for some biodiversity features of concern may result in negative outcomes for other biodiversity features. For example, due to predator–prey relationships, regulations enhancing or intended to enhance the abundance of a threatened emblematic predator may result in reduced biomass of its preys in the OECM area and possibly beyond. If some of these preys are biodiversity features of concern, a difficult trade-off arises, the solution of which may only be resolved case by case, accounting for the unavoidability of natural interactions as well as local and societal relative 'values' of the features concerned.

5.2.3 Flexibility

The need to apply the guiding principles and the identification criteria in a flexible way is stressed in the introduction of Annex III (see also Annex IV, D, f). This flexibility does not mean that it is possible to select what criteria or principles to consider when identifying or managing OECMs. All of them are relevant and should be considered. However, the flexibility is an important provision to enable the design of context-specific regulations that address a range of biodiversity outcome objectives, may be applied in significantly diverse ecological, social, technological and economic contexts, and avoids relying on prescriptive input requirements. Flexibility allows legitimate authorities to adapt and interpret the CBD guidance to their own implementation conditions. It also increases the risk of different interpretations emerging in different countries' jurisdictions and sectors, potentially undermining effectiveness and credibility in cases when the interpretations are sufficiently different. Experience with the criteria is already growing, lessons are being learned about the effects of different interpretations, and best practices may emerge with time. Until such 'best practices' are supported by both the fishing sector and biodiversity conservationists, the guidelines produced nationally or by sectors to better guide implementation of Decision 14/8 (including this present document) are necessarily interpretative but cannot override the Decision itself.

5.2.4 All criteria are relevant and should be considered

All the guidance available in Decision 14/8 is relevant and, in particular, all criteria for identification must be considered and discussed, and the conclusion of the discussions should be registered and archived. Only criterion (A) (not already formally designated as an MPA) is binary and, if not met, is enough to disqualify the area. All others are inherently relative and thus open to a range of responses. In addition, criteria are connected and it is most probably impossible to meet criterion C on biodiversity-related achievements without meeting criterion B on management. Similarly, while paying due attention to criteria B and C, it is important to avoid paying too little attention to the importance of the complex set of other relevant values highlighted in criterion D.

The overall challenge here is to determine what level of management is sufficient, the number and nature of positive outcomes the ABFM is generating or may generate, and which level of benefit is necessary or sufficient for identification of the area as an OECM. Some characteristics mentioned in the Decision 14/8 guiding principles, such as 'connectivity' and 'representativeness', are desirable qualities of all biodiversity conservation areas, but their absence is not sufficient to disqualify an area if criteria A–D are met – just as their absence is not a justification to exclude an MPA from being reported in global coverage targets, nor is their presence enough to justify OECM status if performance is poor on other criteria or subcriteria.

5.2.5 Multidisciplinary evaluation team

The identification requires multidisciplinary expertise and may benefit from multiple forms of knowledge, in order to assess the fisheries; the ecosystem; the biodiversity attributes of concern in the fishery-OECM; the external threats; the

ecological, social, cultural and economic values, etc. The team composition should be clarified from the onset. It needs to be inclusive, with fishery and biodiversity conservation scientists and other holders of local or Indigenous knowledge, management practitioners and representatives of other relevant perspectives, with appropriate roles in the different phases of the process.

Some of the expertise needed will exist already in the fishery agency. However, collaborations typically will be needed: (i) with agencies in charge of biodiversity, to obtain additional expertise and/or to comply with their jurisdiction on the biodiversity attributes of concern expected to benefit; (ii) with agencies managing other uses of the ecosystems, when such uses must also be managed sustainably to deliver the desired outcomes; (iii) with those capable of assessing and addressing the human dimensions of fisheries, particularly small-scale fisheries, and the specific aspects noted in the Decision (such as cultural and spiritual values, and the social and economic implications of the desired biodiversity outcomes); and (iv) increasingly with other experts in academia and environmental organizations who often also have relevant credentials and knowledge and should be fully involved. In all the above, it is important that the search for 'experts' is not carried out narrowly, as has been the case often in the past; specifically, collaborations are needed with experts who are Indigenous elders, community leaders and other respected voices of Indigenous Peoples and other local communities. This expertise is often needed for many parts of the criteria and subcriteria.

The potential institutional sources of expertise might be determined early in the process, and the relevant people and experts called in as needed for each potential OECM examined. To a large extent, the composition of the team needed to undertake the initial identification is similar to that of the team that will later undertake the monitoring, evaluation and reporting (MER) of the OECM performance. It would be very advisable, and in many cases indispensable, to ensure the full participation if not leadership of the MER in the process from the outset (see Chapter 7).

5.2.6 Broadest possible information and knowledge base

In the case of ABFMs that have been in use for years as part of management plans in modern large-scale fisheries, the preferred source of information is empirical data collected from the area being assessed, including scientific data (from natural and social sciences) and expert views. When local information is not available for some properties relevant to the assessment, empirical information from other 'comparable' areas may be used, where 'comparable' takes into consideration the proximity to the site of concern, the similarity of ecological, socioeconomic and governance conditions of fishery, and of other anthropogenic stresses and their management regimes. In addition, the knowledge of Indigenous Peoples and other local communities is always appropriate in evaluation of OECMs, and as the amount of data and analyses from scientific research and monitoring declines, its importance necessarily increases. Guidance for the appropriate collection and use of such knowledge, in the context of free, prior and informed consent (FPIC) by the knowledge holders, is available from several sources.

For biodiversity benefits not previously quantified for an existing ABFM or for which data are not available, or for a newly implemented OECM, the 'evidence' may be produced *ex ante* through modelling, simulations and other ways to extrapolate from 'comparable' cases elsewhere. Regardless of sources, the evidence available on biodiversity benefits may vary greatly in quality and quantity. Its costs are case specific and depend on the ecosystem; the biodiversity attributes of concern; the monitoring and information collection programmes of both sectoral[1] and biodiversity authorities operating in the area; the complexity of the matter; the precision required; and the periodicity of the performance assessments. A large part of the information needed may already be archived in the current monitoring and evaluation system of the fishery or fishery research centre, and in the archives of potential collaborating agencies. It is therefore advisable to identify early in the process possible sources of information and to establish the co-operation needed to access them. The specific information needed for each case may then be acquired incrementally, as the identification proceeds and specific questions are addressed by the experts.

5.2.7 Accounting for uncertainty in social-ecological systems

Social-ecological systems like fishery sectors are complex systems of interacting resources and people within the ecosystem, and under large-scale driving forces. The result is that even in relatively data-rich systems, the available data, analyses and models provide only a partial understanding of the fisheries, their resources and the ecosystem. Notably, a portion of the variation in the attributes being studied seems to be inherently stochastic or dependent on so many externalities that deterministic modelling is not feasible, at least on the scales relevant to OECM status and performance. As a result, uncertainty will have to be recognized and addressed in the identification of OECMs, assessment of their performance and forecasts of their evolution in a fishery context. The issue of the possible cascading effects resulting from regulations intended to generate positive outcomes for some biodiversity features of concern on other biodiversity features is raised in sections 5.2.2 and 7.4.1.

This recognition of uncertainty and the need for risk assessment and precaution in decision making have been present for a very long time, in traditional community-based management (Johannes, 1978, 1982). They became established as necessary parts of scientific approaches to environmental and resource management in UNCED (1992), and have been progressively implemented in fisheries along with adaptive approaches (e.g. Hilborn and Walters, 1992) and biodiversity conservation (e.g. Halpern *et al.*, 2005; Keith *et al.*, 2011; Ouananian *et al.*, 2018). As the OECM has both fisheries and biodiversity objectives, hybrid approaches that emerged in both scientific fields might be needed (see Section 5.2.8 on available methodology). Tools such as multi-criteria decision analysis (MCDA) and risk assessment approaches can contribute to addressing these difficulties, but are highly dependent on the availability of necessary information.

Approaches to ecological risk assessment for the effects of fishing (ERAEF) have been thoroughly researched, for example in Smith *et al.* (2007), Hobday *et al.* (2011), Zhou *et al.* (2016), Lin *et al* (2020) and Fisher *et al.* (2023). The challenges posed by inherent uncertainties in OECM assessments will increase as evaluations progress from identifying and prioritizing potential threats, to assessing risks associated with particular fisheries and biodiversity attributes, to estimating the potential of alternative options to mitigate the risks associated with the fisheries, and to the more inclusive assessments of safe ecosystem levels (SELs) and significant adverse impacts (SAIs) (see Section 5.5.3.2).

Given that assessments in social-ecological systems (and in fishery-OECMs) typically have substantial uncertainty (or confidence limits) in some or all of the factors being considered, the assessment of relative risk associated with multiple options would assist decision makers in making the best-informed decision. This would entail the following considerations.

- Identification of the major sources of risk such as current status and medium-term trajectories of key environmental factors of the fishery and other economic activities; risks from uncertainties regarding management effectiveness, security of policies, and changes in risk tolerances of participants in the governance processes.
- Partitioning risk among the elements of the social-ecological system exposed to the risks such as biodiversity components, ecosystem services, fisheries, coastal communities and other sectors.
- Assessment of the risk characteristics, e.g. duration, extent, amplitude, cost, likelihood of occurrence.
- The socioeconomic and environmental risks associated with alternative decisions, including no action.
- The communication of risks to recipients of the advice and other stakeholders.

5.2.8 Methodological considerations

Methodological considerations of relevance to the identification phase (in this Chapter 5) are also largely relevant for the subsequent performance evaluation of identified OECMs when implemented (see Chapter 7). In the following subsections attention is given to the information and multidisciplinary competence needed, the methods available and the issues related to data-limited situations.

5.2.8.1 Information needed

In a few of the best studied and managed fisheries and ecosystems, much of the information needed to test an ABFM against the OECM criteria would be available or could be constructed from national reports and scientific publications, for the experts to proceed with the assessment. More typically, part of the information needed, particularly but not only on non-target species, broader biodiversity, seabed habitats, ecosystem services and broader socioeconomic values, might require additional data collection and analyses. In this case, the sources and methodologies available locally (including manuals, software, etc.) could usefully be identified from the outset. It is also likely that the multidisciplinary team of experts solicited for the assessment process will carry 'their methodology' with them, which can facilitate access to data and analytical methods not routinely used by fisheries assessment experts.

In all areas, local knowledge may be invaluable and all efforts should be made to involve local knowledge holders in the assessments. The knowledge provided by Indigenous Peoples and other local communities becomes increasingly central to OECM assessments as community-based management grows in importance. An abundant literature exists on assessing the subject, such as Garcia *et al.* (2008) for the integrated assessment of small-scale fisheries and Fisher *et al.* (2015) for application of EAF in such fisheries. The issue has also been well reviewed by IPBES, providing good guidance (McElwee *et al.*, 2020).

5.2.8.2 Multidisciplinary competence

The evaluations are also enhanced by multidisciplinary competences to assess effectiveness in its various bioecological and socioeconomic dimensions. Data from inside the OECM and outside it both contribute importantly to evaluating effectiveness, for example (i) to assess the overall benefits for mobile biodiversity features of concern; (ii) to identify 'control areas' that provide insights into how externalities are affecting the larger ecosystem, possibly making objectives more difficult (or easier) to achieve; and (iii) to develop and parameterize simulations to demonstrate or forecast the outcomes inside and outside the OECM area.

Unfortunately, evidence is still far from complete because (i) fishery-OECMs are recent measures; (ii) broad biodiversity conservation was not an objective for most of them until they sought OECM status; (iii) in many areas the relevant data have not been systematically collected and archived, particularly for non-target resources; and (iv) the potential for replication, control and randomization – the three pillars of experimental science – is practically nil (Gilman *et al.*, 2019; Hilborn *et al.*, 2021b).

5.2.8.3.   *Assessment methodology*

The methods used for the initial assessment undertaken for identification of OECMs and the recurrent assessments of their performance in the MER programme (see Chapter 7) will be largely similar, even though the identification tends to be based largely on available information, while performance assessment relates to the specific objectives of the OECM and requires more in-depth analysis of the data collected through monitoring. In both phases, Decision 14/8 criteria are the fundamental references.

OECMs and fishery-OECMs in particular are too few and too recent to have generated much specific methodological consideration. Moreover, few ABFMs have been recurrently assessed specifically for their effectiveness in delivering even fishery-related outcomes,[2] let alone additional biodiversity outcomes. However, scores of methods, conventional or non-conventional, depending or not on fishery data, simple or sophisticated, have been developed in fishery, conservation and social sciences for monitoring and assessing biodiversity components, ecosystem services and broader social and economic values. Guidelines on biodiversity impact assessment have been produced by the CBD (2006). *Ad hoc* and usually case-specific assessments can be found, and systematic reviews have been recently published, for example in Rice *et al.* (2018), Himes-Cornell (2022) and Petza *et al.*

(2023). There is also an abundant literature on strictly protected MPAs' performance and much less on multiple-use MPAs from which some lessons may be drawn (e.g. Pomeroy *et al.*, 2005; Himes, 2007; Fox *et al.*, 2014; Gill *et al.*, 2017). Methods keep evolving as technology improves and different jurisdictions may prefer to use locally developed variants of broader methods, but making efforts to provide an overview of all available methods is both demanding to attempt and quickly out of date as newer methods and jurisdictional preferences continue to emerge.

Complex ecological models may be used for simulations and scenario analyses when the data and the capacity to use them are available (Fulton *et al.*, 2005, 2015; Plagànyi, 2007; Smith *et al.*, 2007; Trenkel *et al.*, 2007; Shin *et al.*, 2010; Zhou *et al.*, 2011; Collie *et al.*, 2016; Bayley and Mogg, 2019). As available data and capacity decline, there is substantial guidance on how to use multiple sources of knowledge, discussion groups, expert views, questionnaires and qualitative indicators in ways that produce outputs sufficiently reliable to include in management advice (Pomeroy *et al.*, 2004, 2005; Fox *et al.*, 2014; Marnevick *et al.*, 2019; Ivanic *et al.*, 2020). These approaches benefit from highly participatory assessment and management systems, where fishery participants and biodiversity experts have experience participating in assessment processes. The growing Theory of Change initiative (Stein and Valters, 2012) is also developing methods for using such knowledge sources in forward-looking applications, and may be useful in assessing aspects of OECM status in data-limited situations. In some cases, the assessment may be comparatively simple; for example, if bottom-contacting fishing has been banned, it can be argued that the biodiversity benefit is obtained, as long as the measure is effectively enforced. However, as with the identification of VMEs after UNGA Resolution 61/105 was adopted, at least basic information on the status of benthos and its vulnerability to fisheries impacts helps to strengthen the case for the assumed benefits (ICES, 2021).

5.2.8.4   *Testing strategies*

Management strategies based on the advice derived from these approaches might be tested, using both qualitative and quantitative methods, modelling, indicators and expert opinion, and management strategy evaluations (MSE) (Smith et al., 2007). An abundant literature is also available on environmental and biodiversity impact assessment in general (e.g. Bagri and Vorhies, 1997; SCBD and NCEA, 2006; CBD, 2012; Watkins et al., 2015; Butsic et al., 2017; Mascia et al., 2017; Larsen et al., 2019) and in fisheries (e.g. Chuenpagdee et al., 2003; FAO, 2009a; Coll et al., 2014; Langlois et al., 2014).

5.2.8.5 Data-limited situations

In areas with limited assessment capacity and scarce data, expert judgement, possibly using Delphi techniques, and local knowledge (see Coll et al., 2014) might represent a reasonable method for performance assessment, until empirical evidence becomes available (see for examples Himes-Cornell et al., 2022; Petza et al., 2023). The literature on similar areas in similar conditions can be a source of information for some aspects of these evaluations, keeping in mind that unless carefully designed, paired areas are never perfectly matched for comparative purposes. Nevertheless, in some cases fishery reserves and broad habitat protection areas like VMEs can contribute to evaluating 'effectiveness' (Rice et al., 2018; Himes-Cornell et al., 2022; Garcia et al., 2024).

and management capacities and processes acting on more local or sectoral scales. Decision 14/8 does contain qualitative elements of 'proof'. For example, it stresses that biodiversity values should be *important* and *complementary*, ecosystem services must be *essential*, OECMs must be *effective and* the changes in the indicators should be *positive and sustained*. All these terms are qualitative, but specify the features of the properties and the directions of changes needed to satisfy the Decision. The Decision does suggest using *existing standard s... and indicators*, but also *to develop clear, reliable and measurable indicators for assessing the effectiveness of the protected/conserved areas* (in Annex IV, p.17) It also refers to the *degree of protection that the measure offers to the biodiversity components of high priority*. The Decision does not indicate how such 'degree' would be defined and measured (Annex IV, p.19), but makes clear that the degree of protection is a consideration in case-by-case decisions that needs to be estimated somehow and reported in a manner that other interests in the fishery and biodiversity sectors find informative and credible.

The implication for implementation is that the assessment will use the best information available, be it quantitative or qualitative and the level of 'performance' to be obtained and demonstrated is left to the appreciation of the State, within the spirit of the Decision.

5.2.9 Burden and level of proof

Identification of an OECM is strongly based on providing evidence regarding (i) the biodiversity values and ecosystem services that benefit or are expected to benefit from the OECM measures; (ii) the type of governance and quality of the management regulations; and (iii) the probability for both to be sustained in the long term.

The burden of proof for each of these requirements is obviously on the State and/or the legitimate authority in charge of the fishery-OECM. They have the responsibility to prove that they are complying with the Decision, with all the flexibility this Decision allows. How much 'proof' is sufficient is not specified, as is common when a global or large regional decision has to be implemented under the range of assessment

5.2.10 Comparable importance

Principle (d) refers to the need for OECMs to have *biodiversity outcomes of comparable importance ... with those of protected areas*. Obviously, biodiversity outcomes of OECMs and PAs can always be 'compared' and this does not imply that they should be equal in all properties. Furthermore, it is not clear whether having OECMs outcomes comparably important to those of MPAs is to be assessed within a given ecosystem, a country or globally. Does one compare the OECM to the nearest MPA or in the context of the MPAs and OECM network in a given bioregion? However, the aspiration to obtain comparably important outcomes underscores the role that the CBD Parties expect from OECMs individually and in conservation networks. It also indicates that with comparably important outcomes, States and other participants in governance should value OECMs as an approach to conservation as much as they value MPAs. This is both an acknowledgement of the ability of fishery authorities to effectively deliver positive biodiversity outcomes and a warning to fisheries authorities that they need to satisfy more than just other fishery interests that the OECMs are delivering (or can deliver) important conservation benefits.

This principle also suggests that the indicators on which MPAs' contributions to biodiversity are appraised, including in relation to representativeness, relative coverage, ecosystem services and integration/connectivity in larger networks, should be kept in mind when assessing OECM contributions. Since sectoral OECMs are not MPAs, and need (and often do not have) biodiversity protection as the first priority objective (but VMEs do), it is fair to expect that OECM outcomes will complement – but rarely be identical to – well-managed PA outcomes. There can be greater tolerance for activities that negatively impact biodiversity components, as long as those activities, individually or collectively across uses, do not significantly degrade those components. In addition, the nature of the outcomes of any type of 'conservation area' depends on their degree of protection, whether they are strict reserves, multi-use MPAs, sectoral or cross-sectoral OECMs, and whether they are efficiently managed or 'paper parks'.

Some of the biodiversity outcomes relate to components of ecosystem structure and function, and more integrated properties of species, habitats and ecosystems, such as representativeness, provision of specific ecosystem services and degree of integration into larger-scale conservation frameworks. As described in Section 2.3.5.4, quantifying the importance of many of these properties is challenging in its own right, making the demonstration of 'comparable importance' even more difficult. OECMs, like protected areas, will probably result in a range of outcomes, and it is up to the legitimate authority to decide whether the 'importance' of these outcomes for a particular OECM is sufficient to, first, grant the OECM status and defend their decision from potential critics, and then to maintain it in the long term.

5.3 Step 1 – Consolidating Information

In this section, the term 'information' is used in the broadest sense, including raw, organized or processed data (the numbers); grey and scientific literature (reports, articles or books); media (internet sites); and knowledge available from fishers, managers and other experts in multiple disciplines, and holders of Indigenous and Local Knowledge (ILK), necessary and made available to undertake an OECM identification.

This preliminary step is not explicit in Decision 14/8 but is a necessary prerequisite of a more complex process of assessment. In order to facilitate the subsequent steps of the assessment, this step clarifies: (i) the type and amount of information already available for the area; (ii) the sources of such information; (iii) its formats (e.g. paper, digital, in databases or publications); (iv) its degree of accessibility and any constraints on its use (confidential government records of fishery participants, the need for FPIC for sharing of ILK, etc.); (v) the competences needed for the assessment and their potential sources.

The broad types of information needed for each ABFM/area being examined can be deducted from the OECM definition and the list of criteria. They include: (i) the legal status of the area (MPA or not); (ii) its geolocation; (iii) the fishery in which the ABFM applies, other fisheries operating in and around the area, and other sectors potentially impacting the expected biodiversity benefits; (iv) the biodiversity attributes of concern in and around the area, including target and non-target species, protected and iconic species (e.g. Red-Listed species) and the nature of the habitats; (v) the governance, its nature, role, procedures, processes, jurisdictions, legitimate authority; (vi) the fishery management, its area of competence, regulations (spatial and non-spatial), MCS system, illegal fishing (if relevant), ecological and socioeconomic performance, compliance; (vii) ecosystem services provided; (viii) any other locally relevant values including social, economic, cultural or ritual (see Section 5.3.2); (viii) current pressures, potential threats and how they are governed/managed; and (ix) existing conservation networks or seascapes (see more details in subsections below).

Some of this information will probably be recorded on paper or digital form and may be readily collected in advance of the assessment. Some may not be immediately available but might be obtained through collaboration with other institutions (e.g. in charge of biodiversity). Some may be obtained from experts or from ILK knowledge-holders that need to be identified and involved in the process in ways that respect FPIC. Issues of data and information management, archiving and communication are briefly addressed in Section 7.6.

The need to use the broadest possible sources of information is discussed in Section 5.2.6. The importance of consolidating information ahead of the whole multi-stakeholder and multidisciplinary process cannot be overstated. Full consolidation may be easier in well-resourced and structured systems than in data- and capacity-limited areas where not only may there be less relevant data and knowledge, but sources of information may be overlooked because connections among different knowledge sources have not been built or supported. Reports of the ICES-FEG meeting on OECMs in the North Atlantic (ICES, 2021) and in the CDB-FAO-FEG capacity-building meeting in Jamaica (FAO and SCBD, 2023) both illustrate the importance of consolidating information before the meetings that need it are held, and provide examples of how it can be undertaken.

When numerous ABFMs might be considered together (e.g. as in countries with large EEZs and assessment resources), tracing all the information available on all the possible ABFMs may be an overwhelming job, possibly unduly delaying the identification process. As a consequence, the consolidation of information (step 1) and the quick screening (step 2) might be conducted in parallel, quick screening the total set of areas and compiling the information on the most promising ones simultaneously.

Some of the types of information that might be collated are considered in the following sections.

5.3.1 Baselines, thresholds and indicators

It is important to identify pre-agreed reference values (baselines or thresholds) of indicators

to be used for identification of fishery-OECMs and their performance assessments. Although Decision 14/8 presents all the criteria without ranking, in specific cases the legitimate authority may wish to intentionally up-weight or down-weight some criterion to deal with special local pressures and social, economic or environmental circumstances. In such cases, it might be possible to establish the priorities and relative weights from the outset. However, the need for such prioritizations and rankings may emerge from the assessment itself, so agreement on weightings would need to be developed in-stream during the assessment process. In all cases, any proposed prioritizations and weightings and their rationales will need to be submitted to the legitimate authority for discussion and endorsement.

5.3.2 Biodiversity and other relevant values

Consistent with the Preamble to the CBD Convention, Decision 14/8 advises consideration of the full range of ecological, cultural, spiritual, socioeconomic and other regionally or locally relevant values (criterion C1). These values can be identified through a participatory process, calling particular attention to those that could be compromised or reinforced through the use of a fishery-OECM. Scale needs to be taken into account, because some values may be of low priority locally but of high priority at larger scales, and vice versa. The importance of accounting for these values and incorporating them into management discussions is already acknowledged in MPAs and in fisheries management, and the OECM status just increases the context for their use.

The Decision (Annex III, A, a) also refers to the 'significant biodiversity value' expected in OECMs, without further guidance on the level of 'significance'. This issue is examined in detail in Section 2.3.4.1 in reference to principle (a).

Examples of benefits that fisheries-OECMs may provide include the elimination or reduction of fisheries impacts on biodiversity attributes of concern such as: (i) protected, endangered or threatened (PET) species; (ii) spawning grounds and seasonal aggregation of spawners; (iii) concentrations of juveniles; (iv) migration corridors; and (v) important habitats, particularly biogenic habitats. Overlap of fishery-OECMs with areas with high ecological significance like ecologically or biologically significant marine areas (EBSAs), key biodiversity areas (KBAs), vulnerable marine ecosystems (VMEs), particularly sensitive sea areas (PSSAs) or important marine mammal areas (IMMAs) would underscore their biodiversity values.

5.3.3 Information on the ecosystem

The ecosystem(s) within which the fishery-OECM operates should be identified to evaluate its major properties, and to get a sense of interactions between its outcomes and those of other area-based conservation tools in that ecosystem (e.g. in terms of synergy, complementarity and connectivity). This would also help to identify proxies, in similar ecosystems, for information missing in the area being considered as a potential OECM for comparative purposes. Elements of importance for each relevant ecosystem would include: (i) key biodiversity attributes; (ii) the other area-based management measures (e.g. MPAs) with which the OECM might develop synergy; (iii) the specific fisheries within which the OECM would operate; (iv) any other fisheries with which its contribution to biodiversity conservation might have to be integrated; (v) any other sectors whose activities might impede (or facilitate) the intended biodiversity benefits; and (vi) the seascape or other regional framework within which the fishery operates and the OECM might have to be integrated (see Section 6.4).

5.3.4 Information on governance

It is important to establish that the governance processes are likely to be sustained in the long term, and to continue to support the measures providing the biodiversity benefits, regardless of other developments in the fishery or the general area. This important requirement might be satisfied by formal commitments, e.g. in fisheries conservation policies, Fisheries Acts and official communications. The governance processes must also be adequately participatory and inclusive, as discussed in Section 2.3.5.

5.3.5 Expected outcomes

The outcomes expected from a fishery-OECM when it is identified may be already observed (e.g. within the existing ABFM) or expected to occur in a foreseeable future.[3] For existing ABFMs, demonstrating *actual biodiversity outcomes* should be possible if relevant empirical local or scientific information has already been collected and analysed. For existing or newly planned ABFMs in which information on outcomes is not available yet, the claim about *future outcomes* may be supported by several types of information, including anticipatory commitments such as: (i) formal statements of the legitimate authority (policies and strategies) regarding the intended biodiversity outcomes; and (ii) explicit objectives, targets and regulations in the FMP and other marine spatial plan for the area, consistent with an ecosystem approach,[4] and accounting for current or reasonably foreseeable threats.[5] The credibility and strength of such commitments might be judged at least in part by how consistently and effectively the legitimate authority has carried through on statements and objectives in other comparable contexts.

Additional support can be provided by comparative information including (i) information from other areas in which comparable biodiversity outcomes were obtained under similar ecological and environmental conditions with similar measures; and (ii) results of ecosystem modelling management strategy evaluations (MSE)[6] (Punt and Ralston, 2007; Holland, 2010; Ditchmond *et al.*, 2013). The relative strength of these types of 'evidence' depends on the comparability of areas for (i) and the degree of validation and robustness of the ecosystem models and MSEs. The expectation of realizing the desired outcomes would also gain strength if a recurrent monitoring and evaluation system was in place to verify the intended outcomes during implementation, and rules for adaptive adjustments of the measures were also already agreed.

5.4 Step 2 - Quick Screening to Determine Eligibility

The purpose of this step is to increase the efficiency of the identification process, avoiding losing time and resources on ABFMs with low probability to be identified as fishery-OECMs.

Decision 14/8 does not have provisions that specify the process by which a legitimate authority needs to decide what areas to consider as potential OECMs, leaving this operational responsibility to the authority. For fishery-ABFMs, a preliminary screening of the existing ones can increase efficiency of the overall process by both (i) identifying which ABFMs appear more likely to meet OECM criteria and result in a positive decision from a full assessment ('low-hanging fruits'); and (ii) identifying the amount and types of information available to evaluate ABFMs' properties relative to the criteria, and the additional information that would be necessary to allow a conclusive full assessment.

Although many of the same types of information would be considered in both the quick screening and full assessment steps, the results of this first step will help prepare and greatly facilitate the next. In theory, the quick screening may also require a smaller team of experts with broad knowledge of the human and ecological dimensions of the fisheries and key biodiversity features likely to be important in the particular case. This has been demonstrated in the various meetings conducted to explain the OECM concept and the identification process, in which some potential OECMs were rapidly identified (e.g. ICES, 2021; FAO and SCBD, 2023). In practice, however, legitimate authorities might also consider it preferable to set up from the outset a complete multidisciplinary and participative assessment team with appropriate participation or co-operation. Nonetheless, the composition of the assessment teams may vary between a country-wide comprehensive assessment[7] of a large number of ABFMs and the local assessment of a specific ABFM pilot site, for which local detailed knowledge is essential.

When undertaking a comprehensive approach to the assessment, all the places where ABFMs[8] are in use can initially be considered in the identification process. However, only a small proportion of them are likely to satisfactorily meet the OECM criteria, as shown by Petza *et al.* (2019) in Greece, and by Aften and Fuller (2019) in Canada. A preliminary quick screening considering the ABFMs' properties against basic and/or mandatory properties of OECMs, without in-depth analysis, should postpone or remove from the assessment process the ABFMs less likely to be positively assessed, until the weaknesses identified in the first screening are addressed. This focuses available resources on the full assessment of ABFMs with the highest potential to meet OECM criteria. Similarly, when considering an incremental approach, the most promising areas to be considered as 'pilot' areas are likely to emerge from a quick screening process.

The quick screening would check the information available for each criterion and subcriterion listed in Appendix 2. It could stop when it is clear that (i) a mandatory criterion is not fulfilled; or (2) key information to evaluate some criteria is unavailable. For ABFMs that do not hit either of those barriers, it could identify key issues to be considered in the full assessment, identify additional useful sources of information and highlight where additional expertise is needed.

The quick screening process therefore follows the same pathway as the full assessment (criterion by criterion) but in much less detail, proceeding rapidly through criteria A to C looking for a likelihood that the ABFM, under a deeper assessment, would probably meet them. Failing to meet criterion A (not being an MPA) is sufficient to not proceed further. If criterion A and subcriterion B1 (geographically defined) are met, the fuller assessment can be prepared, quickly screening the requirements of the other criteria, identifying the information needed, its availability, and the relevant sources of information. For subcriteria B2 and B3 (on governance, management and current and future threats) and subcriteria C1 to C3 (on effective and long-term conservation of biodiversity values including ecosystem services), it would be sufficient to determine if there are any major inconsistencies or shortcomings relative to any criteria, and if there is enough information to warrant a fuller assessment.

The comprehensive and rich sets of information collated in Step 1 (see Section 5.3) will greatly facilitate the quick screening process. Indeed, if the experts involved in the screening are well informed about the requirements of OECMs, steps 1 and 2 could be undertaken interactively, with simply the amount of information found or missing often pointing to the screening conclusion.

5.5 Step 3 – Full Assessment (See Also Chapter 8)

In this section, the full assessment is conducted criterion by criterion, following their order in Decision 14/8. References are frequently made to the corresponding criteria (e.g. A, B), subcriteria (e.g. B2, C1) and some of the possible elements of evidence (e.g. referred in subtitles as 'Evidence' B2a, C1b). The full list is available in Appendix 2. The structure and discussion here closely parallel that of Chapter 2. Whereas Chapter 2 describes what is required in Decision 14/8, this chapter describes in some detail how to conduct the OECM assessment.

5.5.1 Criterion A: The area is not currently recognized as a protected area

Criteria A indicates that (a) *the area is not currently recognized or reported as a protected area or part of a protected area; and (b) it may have been established for another function.*

The purpose of criterion A is not explicitly stated in the Decision but could be: (i) to avoid adding confusion in the global MPA database which already contains areas of uncertain status, and possible further divergence between national and international MPA databases (see for example Spalding *et al.*, 2016); and (ii) to avoid double counting when assessing the global coverage of these areas in international instruments such as the GBF and the Sustainable Development Goals (SDGs).

5.5.1.1 *Evidence Aa: The area is not an MPA*

The criterion only requires determining whether the ABFM has already been designated as an

MPA, is part of an MPA or overlaps in part with an MPA. It does not call for the area to be assessed against any criteria or definitions of an MPA. If some or all of the area under consideration is already designated as an MPA, that area can be reported as an MPA in the WCMC protected areas database and accounted for in global area coverage figures.

Criterion A can be addressed simply by clarifying the legal status of the area with the legitimate authority, ensuring that it has not been legally designated as an MPA. One should also check whether the area is listed as an MPA in the WCMC World Database on Protected Areas (WCMC-WDPA), and hence potentially used already in the global coverage accounting.

When dealing with existing ABFMs, established under the aegis of a fishery authority, the risk that the ABFM is already designated as an MPA is rather remote, although the risk may exist for totally closed ABFMs dedicated to biodiversity conservation. Processes for legally establishing MPAs usually require at least consultation, and engagement, of the authorities managing many economic sectors likely to beimpacted by the MPA designation, making this risk very low.

5.5.1.2 *Evidence Ab: The area may have been established for other purposes (Ab)*

This statement clearly states that in OECMs, the conservation of biodiversity attributes is an important and necessary objective which, however, may not be the primary objective. The majority of ABFMs have had, historically, the conventional sustainability of the fishery (target stock maintenance and economic viability) as their primary and often sole objective. However, to be considered as OECMs, ABFMs have to produce, in addition, significant positive biodiversity outcomes (benefits). During the last two and half decades, however, with the growing development of the EAF, ABFMs have increasingly taken on explicit biodiversity conservation objectives, sometimes even as primary objectives. Objectives, both primary and secondary, are addressed in more detail in sections 2.3.4.1 Principle (a) and 5.5.2 (Criterion B).

5.5.2 Criterion B: The area is governed and managed

Criterion B is subdivided in three subcriteria: the area is a geographically defined space (B1); the area has legitimate governance authorities (B2); and the area is managed (B3). Each subcriterion is accompanied by suggestions of some elements of evidence to considered. These are explained and interpreted in Section 2.4.2.

Governance and management are central aspects of OECMs and the full assessment would have to evaluate a number of factors, including legitimacy, diversity, equity, collaboration in governance, and how the management systems address participation, effectiveness, ecosystem approach, threats, integration, long-term intent, information, performance evaluation, and cultural and other values.

Assessing all these factors may be complex, particularly in multispecies multigear fisheries, under multiple jurisdictions, and requires a significant amount of information and competence on the fishery and on biodiversity frameworks.

These subcriteria and the elements of information they require are addressed in more detail and in that order in the following sections.

5.5.2.1 Subcriterion B1: The area is a geographically defined space

The information that must be reviewed and reported for B1 has to *describe the area including in three dimensions where necessary* (Evidence B1a). At this point in time, the CBD targets and guidance are silent about the possible vertical zoning of spatial conservation regulations in aquatic ecosystems into pelagic, benthic and possibly one or mor mid-water strata in which different regulations might apply (see Section 2.4.2.3). Certainly, there are both target species of specific fisheries and biodiversity features with distributions largely restricted to only one or a few of the strata and some spatial conservation regulations could be applied effectively in only a part of the water column. However, there are arguments both for and against recognizing such zonation in OECMs (or MPAs). The IUCN-WCPA guidance for both MPAs and OECMs is against such zoning, for ecological and enforcement considerations.[9] However, the need for flexibility and adaptation

to local conditions recognized in Decision 14/8, can be argued to allow States and legitimate authorities to decide, case by case, on how best to regulate the water column, taking both enforceability and vertical ecological connectivity into account. Until there is clarification from future CBD Decisions or COP-approved guidance documents on the issue of vertical stratification of OECM (and MPA) status, unless the regulated depth is defined in the OECM description, it will presumably be assumed that it includes the whole water column.

The vertical dimension of marine OECMs is important as much of the biodiversity relevant to criterion D – on ecosystem services and other locally relevant values – may reside in the water column and although vertical ecological connectivity is important, ocean ecosystems can be strongly stratified. Jurisdictions may also be vertically stratified, particularly in areas beyond national jurisdiction (ABNJ), e.g. outside EEZs and over extended continental shelves, and the control of economic activities occurring at different depth in the same place can be difficult.

Moreover, *the boundaries must be geographically delineated* (Evidence B1b). Thus far, the first fishery-OECMs identified (by NEAF and NAFO), like existing MPAs, have been given static boundaries, whether at shallow depths or in the deep-sea bottom. Geographically defining the space covered by a fishery-OECM involves reviewing both the boundaries set by the legitimate authority for the ABFM management measures in place (indicating legitimate governance) and the spatial occurrence of the biodiversity of concern, including habitat, expected to receive benefits from the OECM (indicating the protected features). The OECM geographic delineation does *not* have to include the full distribution of all biodiversity features of concern as long as its area and the regulations applied in it are sufficient to provide the expected benefits. This means that the initial delineation for B1 early in the assessment may be adjusted after later criteria are assessed, to ensure the area delineated is adequately located and sufficient (see Chapter 8 on revisions).

When the concept of OECMs was added to Aichi Target 11, all discussion focused on static area-based management measures. Conservation of inherently mobile biodiversity features would be sought by combining other species-based management measures with area-based measures to ensure that specific places important to their life histories would be protected. How best to conserve 'places' that were identifiable but mobile was not discussed, possibly because they are rare in terrestrial systems. In marine systems, such 'places' are more common, such as upwelling areas, fronts, gyres and ice fields, and these are often important for biodiversity. At this time, there are no special provisions in Decision 14/8 and its implementation guidance to explicitly accommodate OECMs in dynamic or mobile ocean oceanographic features.

Until the CBD adopts Decisions about how such mobile features are to be treated, the legitimate authority has full scope to decide what geographic boundaries to apply, but the status of any mobile OECM is likely to be challenged by some interest groups as inconsistent with the intent of Aichi Target 11 and KM-GBF Target 3.

5.5.2.2 Subcriterion B2: Legitimate authorities are in place

This subcriterion seeks to clarify whether: (i) the area has a legitimate authority (Evidence B2a); (ii) traditional governance is recognized and empowered, where relevant (Evidence B2b); (iii) governance can be considered 'equitable' in line with the CBD Convention and Decision 14/8's guidance (Evidence B2c); and (iv) governance involves only one or many authorities (Evidence B2d).

These concerns are examined below.

EVIDENCE B2A: GOVERNANCE HAS LEGITIMATE AUTHORITY.   The concern is that the institution dealing with OECMs should have the formal power and means needed to achieve *in situ* conservation of biodiversity within the area; to decide on identification, conservation objectives and management options; to exert oversight; and to report to CBD and WCMC, directly or through the State as most appropriate.

In fisheries, the legitimate authority entitled to establish and manage area-based fisheries management measures (ABFMs or closed areas), to optimize the fishery or reduce its collateral impact on non-target species and habitats, has always been the fishery authority. As an integral

part of the fishery management plan and toolkit, OECMs would be evaluated according to the existing national and international fishery governance and management frameworks, including monitoring and evaluation (see Chapter 7), and control and surveillance capacity. If review of the fisheries management framework found that the fisheries authority lacked authority to conduct any of the listed governance responsibilities in its management of fisheries, those shortcomings would usually carry over to its management of fishing specifically in the OECM, reducing the likelihood that subcriterion B2 would be met. However, having the authority to manage fisheries to deliver harvest management objectives might not be sufficient to provide Evidence B2a, if the fisheries authority lacked authorization or capacity to manage by-catches, habitat impacts of gears, or other ways in which fisheries could affect biodiversity. This deficiency does not necessarily mean the area would fail on subcriterion B2, if it collaborated with authorities managing other sectors whose activities may impact the OECM, including the authority in charge of conservation. Evidence of such collaborations would come from cross-sectoral, spatially integrated policy frameworks, nesting the local to the regional levels, e.g. between RFMOs and Regional Seas Conventions (RSCs).

EVIDENCE B2B: GOVERNANCE BY INDIGENOUS PEOPLES AND LOCAL COMMUNITIES.   This subcriterion states that *the governance by Indigenous People and local communities is self-identified in accordance with national legislation and applicable international obligations* (B2b). The expression 'Indigenous Peoples and local communities' is mentioned 36 times in Decision 14/8,[10] underlining the importance attached by CBD parties to this issue. The concern is to ensure a full participation in identification and governance of the fishery-OECM by the communities depending on the area (the OECM and the fisheries in and around it) for their livelihood, and hence most likely to be impacted by decisions. Key issues relate to self-identification, legitimate authority, and free and prior informed consent (FPIC).

Governance by Indigenous Peoples is self-identified in accordance with national legislation and applicable international obligations. In evaluating the 'legitimacy' of that governance, there would have to be some documentation stating that national policies or legislation devolved management to these stakeholders. If authority is devolved, the devolution would have to respect the equity considerations adopted in the CBD Convention Article 8(j)[11] and the provisions of the United Nations Declaration on the Rights of Indigenous Peoples (UNDRIP) (United Nations, 2007).

The requirement to respect FPIC would be evaluated by (i) any provisions in the policies devolving governance to Indigenous Peoples and other local communities; (ii) commentary from those communities actually allowed to give, withhold or withdraw consent to a project that may affect them or their territories, and to negotiate the conditions under which the project will be designed, implemented, monitored and evaluated. This Evidence is another case in which there is no single 'right' standard for judging performance.

The FPIC principle raises numerous questions regarding implementation, such as: Which person or institution is entitled to provide consent for the community? Would community consent override individual rights of non-community members? How to resolve conflicts? Who are the best information providers? What documentation should be provided to the community? How far in advance? In what language? How can full awareness of the community be ensured? How can existing imbalances in power structures be addressed? How can local knowledge be used for communities' benefits? What appeal mechanisms may exist in case of violation of FPIC? (United Nations, 2005).

When, in the OECM identification process, should the FPIC be obtained? This is not specified in Decision 14/8. The issue of what constitutes 'free, prior and informed consent' and how broadly it applies to the knowledge and contributions of individuals and communities to processes such as OECM evaluations is both sensitive and still developing. Practice currently varies among States. In any particular cases, legitimate authorities should check with national and, as appropriate, local or regional government agencies with jurisdiction over setting standards for knowledge-gathering activities and for rights of Indigenous Peoples and local communities, and ensure compliance with them. Where national or local guidance on FPIC is not available, the evolving UN standards guidance on FPIC should be consulted.[12]

The identification process starts logically with a compilation of the information available for the assessment and this would include both scientific and local knowledge. The latter would imply that some FPIC is needed at the earliest possible time in the process, i.e. when compiling information and for quick screening. However, in South Africa, it has been suggested to seek for FPIC for the specific site-level full identification after the quick screening process has been completed (Marnewick *et al.*, 2019, 2020). Obviously, legitimate authorities may decide what is operationally most appropriate in each case. In most countries where community-managed areas exist, some process may already be in place for fisheries management within which this consultation could take place. For example, in Canada (Government of Canada, 2022) it has been established *how Indigenous information is obtained and its composition, use, and storage should be determined by Indigenous peoples themselves.* While FPIC is mandatory only in the case of areas under the mandate of Indigenous Peoples and other local communities, early involvement of stakeholders and rights-holders from any local community directly impacted by the OECM identification would be strongly advisable.

EVIDENCE B2C: GOVERNANCE REFLECTS EQUITY. The evidence suggested that *governance reflects the equity considerations adopted in the Convention.* The basis for evaluating the equity considerations under criterion B is to assess how costs and benefits of establishing and maintaining the OECM are distributed among all the participants in governance, including levels of government from global to local; corporate and business interests; all potential participants in the fisheries of all scales affected by the OECM; and all other sectors of civil society, including Indigenous Peoples and other local communities potentially affected by the OECM.

Equity has multiple definitions and can be challenging to measure as an outcome. There are social science and economics methods for such measurement, but these take a narrower perspective than does Decision 14/8. Equity requires evaluating (i) formal recognition of rights, identities and values of the people concerned; (ii) inclusive procedures for communication, participation and decision making, which, together, are assumed to lead to (iii) an equitable distribution of costs and benefits among actors, and is presented as facilitating decisions that are taken and implemented legitimately, competently, inclusively, fairly, with a sense of vision, accountably, and respecting rights (from CBD Decision 14/8, Annex II, B, 8). Given these multiple dimensions of equity, this element of evidence requires particular attention in order to quantify the relative degree of success in achieving equity. It does seem clear, however, that evidence that serious inequities are present would be a sufficient argument to fail on subcriterion B2.

EVIDENCE B2D: SINGLE AUTHORITY VERSUS COLLABORATIVE GOVERNANCE. This element of evidence indicates that *the governance may be by a single authority and/or organization or through collaboration among relevant authorities and provides the ability to address threats collectively.*

Collaboration among authorities and organizations whose decisions and actions affect either or both the fishery and biodiversity in the area will be important, and complex for cross-sectoral, transboundary or high seas fishery-OECMs. This collaboration may be documented by Letters of Intent (LOIs), Memoranda of Understanding (MOUs) or comparable institutional agreements. Judgements that the collaborations would help to address both current and future threats collectively and effectively will be subjective, but as with assessing equity, the absence of a history of effective collaboration, or of collaborative agreements to manage possibly serious threats, would be a reason for failure on this subcriterion.

RFMOs and RSOs already offer functional collaborative frameworks, and relationships between these two types of organizations are growing (e.g. in the north-east Atlantic, the Baltic Sea and the Mediterranean). In the north-east Atlantic in particular, the RFMOs (NEAFC, NAFO) and RSOs (OSPAR, HELCOM) all have access to the same international scientific organization (ICES) to obtain assessments and advice regarding both fisheries and the

environment (ICES, 2021), therefore facilitating the assessment of other human impacts in the area, which cross-sectoral co-operation could help address in a more coherent manner.

5.5.2.3 Subcriterion B3: The OECMs are managed

This subcriterion intends to clarify whether: (i) *the area is managed in ways that achieve positive and sustained outcomes for the conservation of biological diversity* (Evidence B3a); (ii) *relevant authorities and stakeholders are identified and involved in management* (Evidence B3b); (iii) *an effective management system is in place* (Evidence B3c); (iv) *management is consistent with the ecosystem approach, including the ability to manage a new threat* (Evidence B3d).

EVIDENCE B3A: ADEQUATE MANAGEMENT. The evidence suggested is that *management is set up in ways that intend to achieve positive and sustained outcomes.* This consideration is practically identical to Evidence (C1a) considered in Section 2.4.3.1 which states that the area 'achieves or is intended to achieve' the expected outcomes. However, Evidence B3a is action oriented and stresses the ways in which the potential OECM is managed to achieve the fisheries management and biodiversity conservation objectives, whereas Evidence C1a is outcome oriented and looks at whether and how effectively the desired biodiversity outcomes are actually being delivered. Both Evidences relate to management efficiency and effectiveness, evaluated from different perspectives.

In order to provide Evidence B3a, it is crucial to realize that an area established long ago, maintained for years or centuries by an Indigenous group, local community or fishing organization, and which remains in place today, may well be part of an effective management system. However, it may not be a management system familiar to 'western' or 'modern' fishery managers. Accordingly, it is important to allow for appropriate evidence of suitable management, rather than a 'one-size-fits-all' evaluation approach. For example, 'adequate management' may be reflected in the existence of a long-standing area together with local knowledge providing evidence of sustained biodiversity.

On the other hand, an analytical approach of the assessment team can involve an inventory of objectives and specific regulations in place or planned in the OECM, such as access rules, gear controls, authorized practices, control and surveillance, and their expected outcomes. For this purpose, jurisdictions with adequate expertise and technical capacity, using, for example, tools like MSEs, can undertake very rigorous evaluations of whether the management plan is likely to achieve the fishery outcomes. Conceptually, nothing prevents MSE approaches from also evaluating the likelihood of delivering associated biodiversity outcomes, but such extensions would require substantial data or information on all the major threats the fishery poses to biodiversity, and how effective various management regulations have been at managing those threats. Those are ambitious preconditions even for capacity-rich jurisdictions. Therefore, alternative approaches are needed for this part of the assessment, particularly in areas with limited data and capacity. The information collected and the conclusions reached in the assessment of Evidence B3a and C1a should be shared among the assessors concerned. And the two assessments could indeed be conducted together.

Additional elements of evidence of an adequate management system include: (i) an inventory of measures to address current and future threats; (ii) the existence of contingency plans and decision rules within a precautionary approach; (iii) related improvements of the monitoring and evaluation system; and (iv) the clear integration of the OECM area and functions in the fishery management plan. These elements are examined briefly below.

- *Inventory of regulations to address current and future threats on biodiversity.* The inventory of regulations to address current and future threats is not explicitly required in the Decision but is needed for the assessment, in many places where threats and related actions are mentioned. Such inventory would be accompanied by information on the potential effectiveness of such regulations on the biodiversity attributes of concern in the area, and factors that could enhance or reduce the effectiveness. Such information could be codified, archived across different OECM

assessments and made available for other assessments, including of subcriteria C1 and C2. This subcriterion simply requires that management *has the ability* to achieve the biodiversity outcomes so a strongly quantitative evaluation is not necessary as long as achievement is shown to be feasible, and the use of local and Indigenous knowledge may be crucial for this. Where regulations can be adapted, the evaluation can identify regulations that have both potentially highest return on investment and lowest cost/benefits ratios (see Section 5.5.3.1), as well as the circumstances in the fishery, its governance and, where feasible the ecosystem, that could be adapted to improve effectiveness of the management regulations.

- *Designing contingency plans and decision rules, and applying precaution.* Contingency plans, decision rules and precaution are not explicitly addressed in Decision 14/8 but are implicitly needed for management to address threats properly. This should be done to the fullest extent appropriate (FAO, 1996b), balancing risks of misses and false alarms. Data scarcity and management difficulties call for raising the level of precaution in the risk management framework. Advice could also be developed on ways to integrate the action across fisheries and within the ecosystem (to prepare the integration addressed in Chapter 6).

- *Advising on the improvements eventually needed in the existing MER.* Improving the capacity of the OECM to 'achieve' its objectives may require strengthening the capacity to monitor the biodiversity attributes' evolution in the OECM, and specifically evaluating the performance of the regulations applied to and around it (see Chapter 7 for details) and also to ensure the long-term intent of the OECM. This comment is also relevant when considering criterion C4 on information and monitoring.

- *Integrating the fishery-OECM in the fishery management plan.* The 'management plan' in any given location will depend on the context. It may vary considerably between a self-managed Indigenous fishery, a large-scale fishery or a small-scale fishery in a developing country. Integrating the fishery-OECM into the management plan is accordingly equally context specific. In any case, the original ABFM (e.g. local sacred site or governmental closure) and its specific regulations may already be integrated into the management plan. However, the latter may need updating to incorporate additional biodiversity-related objectives and regulations (as required). In addition, for a positive evaluation that the fishery-OECM is managed in ways that achieve positive and sustained outcomes for conservation of biodiversity, its individual objectives (primary and secondary) and the special technical regulations applying in it (e.g. regulating access, gears and practices) should not only be mentioned in the fishery management plan (FMP) of the fisheries in which it applies, but effectively pursued, enforced, monitored and assessed, adapting the MER system as required. The degree of integration is hard to quantify but, again, the absence or failure of integration of the OECM requirements in the FMP should be apparent, and justify failure on this sub-criterion. The issue is addressed in detail in Chapter 6.

EVIDENCE B3B: INVOLVING OF AUTHORITIES AND STAKEHOLDERS. The Evidence suggested is that *relevant authorities and stakeholders are identified and involved in management.* This evidence requires both identifying who are 'relevant authorities and stakeholders' and documenting that they are 'involved'. The issue has three dimensions: international, national and local. At the international level, the legitimate authority would be the RFMO with competence in the area and for the ABFM under consideration, and as such the Contracting Parties would mandate the Secretariat to submit the identified OECM to WCMC for inclusion in the World Database on OECM. Stakeholders at this level include observer organizations, such as NGOs, academia and industry. At national level, top authorities and actors' representatives, unions and NGOs may be most important. At local/site level, decentralized public staff, local authorities, co-operatives, individual fishers and knowledge-holders are essential. With regard to authorities, minimally the legitimate authority (or its representative)

should have a leadership role, but officers from departments or agencies with fishery or biodiversity responsibilities at all levels of government could be relevant, and should at least have been aware of the State's interest about OECMs, and the specific consideration of the local ABFM as a potential OECM. Beyond that, the complexity of establishing who are 'relevant' stakeholders' was discussed in Section 2.3.5.1 in relation to principle (g) on consultation.

The Evidence should consider the degree to which (i) communities of practice and rights-holder and stakeholder networks are developed and fostered to facilitate mutual learning and exchange, and support governance, monitoring, enforcement, reporting and assessment; (ii) a common understanding has been developed across rights-holders and stakeholders regarding the objectives and expected outcomes of OECMs; (iii) social and communication skills of managers and practitioners of OECMs are effective, and (iv) a process should exist for accommodating new interested groups and perspectives that may emerge over time. Again, the diversity of situations means there is no 'perfect' score, but obvious shortcomings on any of these factors, such as interested actors who claim to be excluded from involvement, are negative considerations with regard to meeting subcriterion B3.

EVIDENCE B3C: A MANAGEMENT SYSTEM IS IN PLACE. Management 'systems' are mentioned only twice in Decision 14/8, with no further detail. As explained in Section 2.4.2, in larger-scale fisheries, a management system should have a number of properties including (i) the legitimate management authority; (ii) the dedicated legislation (e.g. Fisheries Act) and regulations; (iii) supporting services; (iv) the oversight and advisory bodies; (v) the stakeholders; (vi) the network of collaborations, (vii) links to the Navy and coastguards, for control, surveillance, interception and arrest, and the judicial institutions (for trials). Community-based fisheries may not need all this management overhead, but do need at least to have their self-management or co-management formally recognized by the national legitimate authority, and credibility with the community.

Not all the seven features listed for a management system may be equally important to ensure it effectively 'contributes to sustaining in *situ* biodiversity conservation' (as stated in B3d). However, a positive evaluation of this subcriterion would require at least evidence that (i) there is a legitimate authority; (ii) there is at least a unit in charge of management of the fishery; (iii) the fisheries management unit co-ordinates its efforts with a governance unit in charge of biodiversity; (iv) biodiversity-related objectives and possibly targets exist; (v) there are specific regulations in place that can manage the fishery impacts on biodiversity; (vi) the regulations are enforced; and (vii) there is some more or less regular and formal monitoring allowing performance evaluation. Absence or major shortcomings in any of those properties would significantly weaken the evidence that an effective management system is in place, and argue against a conclusion that the ORCM is managed (subcriterion B3).

EVIDENCE B3D: THE ECOSYSTEM APPROACH AND THREATS. The Evidence B3d states that *management [should] be consistent with the ecosystem approach with the ability to adapt to achieve expected biodiversity conservation outcomes, including long-term outcomes, and including the ability to manage a new threat.* The issues have been briefly considered in Section 2.4.2.3d. From a fisheries point of view, the EAF has been formalized in the FAO technical guidelines. It is being progressively implemented and evaluated, to consider both ecological and human dimensions. A key aspect is work towards reducing collateral impacts of fishing, in which fishery-OECMs provide an opportunity to practically improve the EAF implementation.

For adaptive management, even when there are provisions to allow or promote adaptive management, limitations on governance capacity and operational difficulties constrain the ability to anticipate, detect and respond effectively to new emerging threats to biodiversity. Therefore, in case-by-case assessments, if the EAF is not explicitly acknowledged in the FMP, there would need to be provisions that could reasonably be expected to (i) achieve expected biodiversity conservation outcomes, and (ii) respond adaptively to changes for this subcriterion to be met.

With regard to management of new threats, and to support a positive decision on meeting this criterion, evidence would be needed on (i) efforts to identify current or reasonably anticipated threats that affect or will affect biodiversity in the exploited ecosystem; and (ii) that potential OECMs (with the technical regulations taken inside and around them) have a reasonable capacity to reduce or eliminate the related risks.

5.5.3 Criterion C: The area achieves a sustained and effective contribution to *in situ* conservation of biodiversity

Criterion C states that the area *achieves sustained and effective contribution to in situ conservation of biodiversity.* The four subcriteria specify that the OECM is effective (C1); its benefits are sustained over the long term (C2); the area contains important biodiversity attributes (C3); and the OECM is monitored, its performance is assessed, and the archived information is broadly available.

The criterion should be evaluated in a logical sequence. For example, first, identify, what the important biodiversity attributes of concern are in the area (subcriterion C3). Second, assess the state of these attributes; identify the pressures exerted on them and the regulations in place to confirm that the expected benefits are – or are likely to be – achieved (subcriterion C1). Third, confirm that these benefits are – or are likely to be – maintained for the long term (subcriterion C2). And fourth, confirm the existence of a monitoring and information management system for long-term assessment and adaptive management (subcriterion C4). These subcriteria and the evidence they may require are addressed in more detail and in that order below.

5.5.3.1 *Subcriterion C3:* In situ conservation of biodiversity

This subcriterion is entirely dedicated to one important consideration: the *identification of the range of biodiversity attributes for which the site is considered important (e.g. communities of rare, threatened or endangered species*[13]*; representative natural ecosystems; range-restricted species; key biodiversity areas (KBAs); areas providing critical ecosystem functions and services*[14]*; and areas for ecological connectivity.*

To apply this subcriterion, the review would consider elements of biodiversity (e.g. important life stages, species or habitats) present in the OECM, and possibly also outside it,[15] that are: (i) identified by a mandated agency, or widely supported social process, as a conservation priority, e.g. listed as endangered, threatened or protected in national or international legislation, or by relevant Indigenous Peoples and other local communities as culturally iconic; and (ii) impacted by fishing operations and for which conservation regulations are required to eliminate, reduce, mitigate the impact, and eventually restore healthy conditions. If these 'biodiversity attributes of concern' for the fishery were also impacted in the same area by other sectors, this should be documented as much as possible. The resultant inventory in itself does not have to meet any standards other than adequately capturing the main biodiversity features of concern in the area, as judged by experts (either scientists or elders from local communities of Indigenous and other local people.

Meeting this subcriterion does not necessarily require pristine habitats and unaffected populations. It does require that the habitats have all 'natural' features and are not undergoing degradation, and that the populations of the characteristic species are healthy. If species were previously depleted and habitats degraded, regulations should be in place for their recovery. Depleted populations that are not being impacted by the fishery are not sufficient justification to fail on this criterion, but are indicative of pressures that need to be identified and addressed by appropriate regulations applied by appropriate jurisdictions if objectives related to healthy ecosystems are to be achieved.

5.5.3.2 Subcriterion C1: The area is effective

In evaluating this subcriterion, the biodiversity attributes of the potential OECM, particularly the attributes of concern for the fishery sector, have already been listed in the preceding Section 5.5.3.1 (subcriterion C3). Clearly, in this subcriterion, the term 'area' refers to the combination of the area (boundary, location, dimension, etc.) and the regulations applied in it. Evaluating effectiveness would be greatly facilitated if the state and trend of the biodiversity features of

concern were known. Assessing both for every biodiversity feature in the potential OECM may be an unreasonable burden, but when preparing the inventory of 'important biodiversity' for subcriterion C3, if any species or habitat features were identified as of particular priority for any of the reasons specified in this subcriterion (e.g. communities of rare, threatened or endangered species, etc.), something should be known of their status. The elements of evidence that would provide the baselines needed for evaluation are examined below.

EVIDENCE C1A: THE AREA ACHIEVES IN SITU CONSERVATION. This element of evidence states that *the area achieves, or is expected to achieve, positive and sustained outcomes for the in situ conservation of biodiversity.* The specific achievements of the fishery-OECM, or any other set of management regulations, will be different for different biodiversity attributes, for different reasons: (i) trophodynamic relationships such as trophic cascades; (ii) different life histories: short-lived species respond sooner to regulations than long-lived ones; (iii) other ecological factors. Consequently, OECMs cannot be expected to yield positive benefits for all biodiversity features, and the evaluation should focus on the biodiversity features of concern (species populations and habitats) identified above. There should be evidence coming either from the OECM itself or from similar taxa in comparable ecosystems and situations, that the OECM regulations maintain the priority population sizes and habitats when they are healthy, and promote rebuilding if they are depleted or degraded. The biodiversity objectives for the OECM do not have to be set for all the priority species and habitats in the area, but the activities allowed and regulations applied in the fishery-OECM to benefit specific biodiversity features should not facilitate or contribute directly to depletion or degradation of other priority species or habitats of concern, whether or not they are explicitly listed in the OECM objectives.

From the above baselines and evidence of positive (or negative) impacts of the fishery-OECM on the biodiversity features of concern, future effectiveness can only be inferred, taking into account: (i) the likely future changes in nature or intensity of threats; (ii) the likely consequences of these changes in threats on biodiversity attributes and eventual new concerns[16]; (iii) desirable and possible improvements in the management regime of the OECM and, as appropriate, in the fishery, including the additional regulations needed to obtain such improvement. Then, expert knowledge, supported by projection models when available for the area, can be the basis for foresight regarding expected outcomes from the OECMs.

There are many approaches and methods available to explore the nature and likelihood of the future biodiversity outcomes of concern in a potential OECM in both data-rich and data-limited situations. The abundant literature on methods used to assess the impacts of MPAs on biodiversity, for example, is also relevant for OECMs, particularly the literature relating MPAs and fisheries (e.g. Alban *et al.*, 2011; Garcia *et al.*, 2013; Todd *et al.*, 2013; Affllerbach *et al.*, 2014; Weigel *et al.*, 2014; Sciberras *et al.*, 2015; Sadio *et al.*, 2015; Fulton *et al.*, 2015; Russi *et al.*, 2016; Ban *et al.*, 2017; Leite *et al.*, 2019; Ward *et al.*, 2019). When evaluating the positive and sustained outcomes, Decision 14/8 requires *clear benefits to biodiversity conservation* (in Annex III, A, c). It recognizes that such benefits may not be the *primary intended management objective* of the OECM but states that it is desirable that initially unintended co-benefits *become a recognized objective of the OECM management* (in Annex III, C, 1, d).

Regardless of the methods chosen to evaluate how well the fishery-OECM contributes to maintain or enhance biodiversity and to reduce threats to biodiversity components, some benchmarks are needed for determining the effectiveness. CBD Aichi Target 6 (in 2010) referred explicitly to the need to avoid SAIs, *sensu* UNGA Resolution 61/105, and the FAO guidance on SAIs was acknowledged in CBD COP Decision X/29 (paragraph 54).

SAIs are defined as impacts that compromise ecosystem integrity, i.e. ecosystem structure and function, in a manner that: (i) impairs the ability of affected populations to replace themselves; (ii) degrades the long-term natural productivity of habitats; or (iii) causes, on more than a temporary basis, significant loss of species richness, habitat or community types (FAO, 2009). Items needed to identify SAIs would include: (i) the intensity or severity of all impacts from fishing or – to the extent feasible – cumulative impacts from other activities on the area; (ii) the absolute and relative spatial extent of these impacts (compared to the area covered by the biodiversity attributes of concern); (iii) the sensitivity/vulnerability of the biodiversity attributes to the impact(s); (iv) the ability of the component to recover from identified harm (resilience) and the potential rate of such recovery; (v) the likely changes to ecosystem functions given the impacts in items (i) to (iv).

Avoidance of SAIs in a fishery-OECM is, therefore, a minimum standard that must be met. However, just meeting such minimum standard is not sufficient to ensure a high likelihood of 'positive and sustained outcomes for the *in situ* conservation of biodiversity'.

The concept of a safe ecological limit (SEL) is not referred to in Decision 14/8 but it is worth considering in the evaluation of how OECMs contribute to biodiversity conservation. Aichi Target 6 also calls for impacts of fishing on species, stocks and ecosystems to be kept within SELs. Although the concept has never been precisely defined in operational terms, e.g. with clear measurable targets and units for its quantification (Donohue *et al.*, 2016), its origin in the 'Planetary Boundaries' framework (Rice and Garcia, 2019) provides a direct link to well-established fishery stock assessment frameworks. These frameworks have been used for decades as a foundation for contemporary single-species fisheries management, to identify limits for stock status, minimum spawning biomass, stock–recruitment relationships, maximum fishing mortality, and to develop strategies and decision rules to maintain stocks within safe limits. They can be used much more widely in evaluating biodiversity status in the context of positive and sustained biodiversity outcomes, particularly when augmented with consideration of tipping points, regime limits and shifts, factors that are part of the more encompassing Planetary Boundaries framework.

Although challenging to establish even within standard fisheries management frameworks, these additional considerations could be applied for broader biodiversity conservation, defining safe limits below which ecosystem properties (e.g. structure and function) may not be driven. Parametric and non-parametric methods used to define relations and critical inflexion points for safe limits, including in situations of uncertainty, are well known and do not require an understanding of the full functional relationship between the ecosystem state property and the functions it serves (ICES, 2002; Rice, 2009; Cadrin and Dickey-Collas, 2015). Approaches exist to deal with data-rich and data-poor situations (Fulton *et al.*, 2016; Canales *et al.*, 2018). Even in much simpler conventional fisheries stock assessments, simply feeding the best information available into a standardized assessment algorithm does not provide a complete answer to the current health and trajectory of a stock. However, all the knowledge and experience gained by interpreting the results of these methods in fisheries will carry over to their wider application in evaluating the likelihood and potential nature of the expected positive and sustained biodiversity outcomes.

The inventories of the biodiversity features of concern (see Section 5.3) and of the regulations in place (see Section 5.5.2.3, subcriterion B3) in and around the potential fishery-OECM should have already been done. This part of the evaluation considers the relationship between the local impact(s) on biodiversity to be addressed and the expected outcome(s) of the measure(s) being considered. The intention is to assess the effectiveness of the regulations(s) being proposed, reflected in the likelihood and potential magnitude of their outcomes. The effectiveness of the full suite of regulations in the potential OECM should be evaluated as a package, but the evaluation can also consider single regulations or subsets of them, if management has the flexibility to add, drop or adapt regulations to improve the likelihood or magnitude of the expected benefits.

Although direct local evidence of effectiveness is always important, it is not always available. Even when there is local evidence, information on the effectiveness of similar regulations to deliver the desired outcomes in comparable ecological, socioeconomic and governance settings can provide a larger information base for the evaluation. When using evidence, whether local or from other areas, the considerations made in Step 1 (see Section 5.3) help inform how different types of natural and social science evidence and the knowledge of Indigenous Peoples and other local communities can be used and integrated.

In case an upgrading of the ABFM is envisaged to better meet the OECM criteria and support a higher likelihood of delivering better biodiversity outcomes, additional protection measures may be considered, such as (i) adjustment of the area boundaries to improve ecological protection; (ii) joining of neighbouring areas to improve connectivity; (iii) adding technical regulations regarding fishing gears and practices; (iv) increasing penalties for deterrence or incentivizing compliance, etc. Whether considering measures currently in place or additional ones for possible implementation, the costs and feasibility of implementing and maintaining the measure in the longer term should be part of the evaluation.

EVIDENCE C1B: THREATS ARE ADDRESSED EFFECTIVELY. The evidence would be that *threats, existing or reasonably anticipated ones, are addressed effectively by preventing, significantly reducing or eliminating them, and by restoring degraded ecosystems.* Dealing with threats is an important management responsibility, and the subject has been addressed under subcriterion B3 (see ection 5.5.2.6). However, criterion B focuses on governance and management of the negative forces (threats) and the regulations adopted to deal with them, whereas criterion C focuses on the biodiversity attributes affected by these threats and evidence of the protection. These are complementary aspects of how effectively the OECM affects/reduces threats to biodiversity, although individual regulations may emphasize either addressing threats (in criterion B) or improving biodiversity under threat (in criterion C). Effective regulations can do both in complementary ways. For example, an area closed to fishing may both reduce excessive fishing pressure and protect seabed benthos. However, regulations effective at either one or the other of these aspects also contribute to OECM status. For example, altering the area or the timing of a fishery may reduce seabird entanglements in gear, without reducing overall fishing pressure on the system.

The assessment of pressures (and threats) on biodiversity should include their identification, their source, the description of their nature (drivers, factors and mechanisms involved), and the evaluation of their current or potential impact on biodiversity and related social and economic costs if they did occur. As fully as possible, information should also be assembled about actions taken in other sectors or jurisdictions to reduce, mitigate and eliminate threats and pressures to biodiversity in the area of the OECM, possibly with the outcomes of these actions, facilitating cross-sectoral or regional action.

A range of methods is available for the assessment, such as biological and social surveys, time series analysis, community-level case studies (for that scale of OECM) and various other approaches (see Section 5.2.8). The pressure–state–response (PSR) framework (Moldan *et al.*, 1997; Chesson, 2013) and the Multi-Criteria Decision Analysis (MCDA) (Fletcher, 2008) could be used to organize the information and guide a participative assessment. All identified threats should first be examined and their likelihood be at least qualitatively judged (e.g. unlikely, negligible, likely, significant). Rationales for each judgement should be recorded and reviewed periodically to accommodate both increased information and changes in conditions affecting the various threats, such as changing market conditions, new technologies, etc. The evaluations of how effectively other non-negligible threats to biodiversity are managed in the OECM should apply the same considerations as when assessing the effectiveness of the OECM measures presented in sections 2.4.3 and 5.5.2.3. These coarse evaluations of other threats in the same area become part of the final synthesis report (see Section 5.6) and are taken to any discussions of multilateral conservation initiatives.

EVIDENCE C1C: MECHANISMS TO DEAL WITH THREATS. The suggested evidence is that *mechanisms, such as policy frameworks and regulations, are in place to recognize and respond to new threats.*

This Evidence is about the action needed as a new 'current' threat materializes. Threats may originate (i) in the fishing area, due to fisheries or other sectors; or (ii) from outside it, e.g. from land-based pollution, illegal foreign fishing, long-range oceanic connections or climate change (a consequence of atmospheric pollution). For effective management, the source

of the new impacts must be ascertained and the needed collaborations with non-fishery institutions established.

In examining that element of information, attention should focus on the overarching policy frameworks of the legitimate authority for the fishery, and for any other threats evaluated as likely or significant (Evidence C1b). The evaluation should consider the several key factors.

- For each potential threat, is monitoring in place that would be likely to detect new or increased impacts from the pressures imposed by the threat?
- If there was concern about an increasing threat, how rapidly could the authorities managing the threat identify and implement new or adapted regulations to reduce the threat?
- How effectively can the authorities manage the threats, with the management regulations available to them?
- When risk of a future increase in a threat is moderate or high, are contingency plans formulated by the appropriate authorities in place for proactive and precautionary actions, should there be evidence that the threat is increasing?

All of these are qualitative evaluations, drawing from readily available information sources. Evidence C1c should focus on proactively finding major failings on any of the questions above and addressing them, to ensure the positive biodiversity benefits can be sustained in a changing world. If no major shortcomings are identified, this Evidence will strengthen the assessment of subcriterion C1 and criterion C. However, just as the evaluation of likelihood and magnitude of various other threats to biodiversity should be reviewed periodically (see preceding section), the ability and preparedness to address changing threats should be reviewed as well, for the same reasons.

EVIDENCE C1D: MANAGEMENT INTEGRATION. The evidence is that *to the extent relevant and possible, management inside and outside the other effective area-based conservation measure is integrated.* It relates to the likely effectiveness of the potential fishery-OECM. Chapter 7 reviews the opportunities, challenges and benefits of integrating

OECMs: (i) within the specific fishery management plan; (ii) in the whole fishery sector; (iii) across economic sectors in an EEZ; and (iv) at regional level, within seascape or similar frameworks. Consistent with the qualifier 'to the extent relevant and possible…' in the formulation of the Evidence, there has almost always been scope for greater integration of management at each of these scales, so there is no 'perfect score' to seek in assembling this evidence. Rather, the evaluation should highlight opportunities for improving integration on each scale, and point out any threats to effectiveness or efficiency from management regulations and actions that are inadequately integrated.

5.5.3.3 Subcriterion C2: Sustained outcomes

This subcriterion provides that *the OECMs are in place for the long term or are likely to be*, specifying that, in the Decision, '*sustained*' refers to the continuity of governance and management, while '*long term*' applies to continuity of biodiversity outcomes. In practice, the two terms tend to be used as synonyms.

The terms '*sustained*' and '*long term*' often appear together in the OECM definition, in principle (h), subcriterion B3 and criteria C1 and C2, stressing the importance of a *sustained* management effort to maintain the positive outcomes for the *long term*. However, the Decision leaves to States the responsibility for deciding what would constitute an acceptable 'long term' for an OECM. However, considering how much ocean areas and biodiversity features can differ in their inherent variability, and vulnerability to unmanageable externalities such as large oceanographic anomalies, it is impossible for any State to assure that any benefit will persist forever. In addition, ecological trophic cascades will transfer benefits across the food web, generating natural changes in energy flows and, potentially, in biodiversity structures. It is therefore imperative that the 'long-term' concept be understood as flexible and adaptable, to obtain the best possible long-term overall benefits, case by case.

The concern with the perspective of spatial measures being dropped or altered frequently by the legitimate authority to respond to socioeconomic pressures, at the expense of biodiversity, is understandable, justifying the stress in Decision 14/8.

Although the lifespan of an ABFM did not seem to have been an important issue for measures assumed to remain in place 'as long as needed', granting an OECM status to it requires some clear statement and evidence of the long-term intent of the measure, compatible with the interest of the biodiversity features of concern and with the need for dynamic adaptation of the fishery and biodiversity management to medium- and long-term ecological oscillations and trends.

Unless tied to legislation or binding agreements that would make individual management measures (spatial or otherwise) very difficult to alter, evidence on intent to maintain a measure in the 'long term' will be inferential and be included in the documentation used to identify and report an OECM. The history of the actions of the legitimate authority or other agency applying important complementary resources (for example, when surveillance is by a different agency or favourable financing terms are provided by another agency or the private sector) will always be a consideration. For instance, in the NAFO's case, the VME closures identified as OECMs have a 5-year duration and are subject to scientific reassessments and renewal by the Commission (i.e. NAFO's decision-making body). The Annual Meeting of the Commission in 2023 approved the advice received from the NAFO Joint Commission-Scientific Council Working Group on Ecosystem Approach Framework to Fisheries Management (WG-EAFFM),[17], which stated that any changes of management regulations to these OECMs would have to be communicated to the WCMC and the CBD.[18] Since this decision is documented in an official meeting report and has been approved by the Commission, it is reasonable to conclude that there are sufficient safeguards in place to ensure that either the measure will be maintained in the long term/sustained or that if it is not, the OECM submission will be adjusted to reflect any possible amendments to the current management regulations that, together with all the other information, support the identification of the OECMs.

The types of evidence regarding the legitimate authority's willingness to '*sustain*' the management effort for '*long-term*' outcomes could include: (i) evidence of long-term formal policy and legal frames, institutional arrangements (central or local), legal or strong socio-cultural requirements; (ii) a formal statement or commitment from the high-level governance[19]; and (iii) historical record of the existence of the ABFM before its identification as an OECM; this is a crucial aspect for local-level and community-based conservation areas, which may strongly reflect cultural and spiritual values and/or be based on proven benefits to sustainable livelihoods; (iv) by the existence of a functional monitoring and recurrent evaluation system; (v) identification of enabling (or impeding) factors, likely to help maintain (or reduce or cancel) the expected biodiversity benefits or co-benefits of a potential OECM and the regulations applied inside and around it, or from other likely changes to the environment; (vi) assessing the direct and indirect social and economic costs and benefits of the OECM, and their distribution among communities, social groupings and economic interests affected by the measure to evaluate the likelihood of stakeholders' support in the long term; (vii) the provision of financial or other incentives to strengthen support of the regulations by the fisheries, particularly when accompanied by assurances that other regulations in the FMP still give the management sufficient flexibility to cope with changing circumstances, including climate change; (viii) assessing the dependence of the OECM benefits on the conditions outside the potential OECM area, e.g. on the complementary management regulations in place in the fishery(i.e.s) operating in and/ or around the OECM; connectivity with other OECM areas or within the MPA network; or land-based pollution and other impeding factors; and (ix) the governance processes formally involving stakeholders likely to support persistence of the spatial measures in the long term.

The expectation of sustained and long-term biodiversity benefits also raises, implicitly, the problem of the existence or likely occurrence, in the future, of non-fishery threats that fishery regulations would not be able to address. It also raises the issues of risk assessment, monitoring, enforcement and adaptation to new situations (including climate change) as well as revision and revocation procedures, e.g. if and when an OECM is found to no longer satisfy the conditions that led to its identification. Ways to address these concerns are presented in sections 5.2.7 (Accounting for uncertainty), 5.5.2.6 (Inventory of threats) and 5.5.3.7 (Ecosystem functions and services).

5.5.3.4   Subcriterion C4: Information and monitoring

Subcriterion C4 stresses the need to have in place a comprehensive information management and monitoring system to provide the data needed to assess the present and ongoing OECM effectiveness, although provision of *evidence* of such effectiveness is optional when registering the OECM in the world OECM database (see WCMC, 2019). The elements of evidence suggested are: (i) documentation of the known biodiversity attribute,.... relevant, cultural and/or spiritual values..., and the area and the governance and management in place (Evidence C4a); (ii) a monitoring system that informs management on the effectiveness of measures with respect to biodiversity, including the health of ecosystems (Evidence C4b); (iii) processes in place to evaluate the effectiveness of governance and management, including with respect to equity (Evidence C4c); and (iv) availability of the general data of the area such as boundaries, aim and governance (Evidence C4d).

Elements of evidence (C4a) on biodiversity attributes and other values and (C4c) on governance and management have already been considered and eventually collected during the identification process and discussed in other parts of this Section 5.5 and do not need to be addressed again here. The specific elements of evidence are (C4b) about the monitoring system, and (C4d) about the public access to the information.

EVIDENCE C4B: MONITORING SYSTEM. Some form of monitoring system is needed to collect, manage and analyse the information needed to describe the OECMs and the regulations applied therein, and to recurrently assess their performance with respect to biodiversity and ecosystems health. In some fishing nations, such monitoring is already available, using fishery data and scientific surveys, and it may need to be updated to monitor additional biodiversity feature of concern and other elements not previously required for ABFMs. In more traditional management systems, in small-scale fisheries (SSFs) and coastal communities, the local knowledge holders are usually the 'repositories' of such information. As the many provisions of the KM-GBF related to the rights and the knowledge of Indigenous Peoples and other local communities are implemented by States, and as Indigenous and local governance expands through self-management and co-management, methods might be developed to ensure long-term accessibility to and coherence of the information, consistent with the standards and goals of FPIC.

EVIDENCE C4D: INFORMATION AVAILABILITY. All these types of evidence may be made publicly accessible on websites of the legitimate authority and other interested groups, and some of this information is requested for registration of the OECM in the WCMC database (see Section 7.5.2). As with degree of 'integration', there will always be opportunities to increase the amount of information on the OECM, including the biodiversity in the area and its performance over time. Moreover, this subcriterion refers to the necessity for information management systems, on fisheries and on biodiversity and such systems are continuously improving and expanding. Consequently, there will never be an upper bound to the amount of monitoring undertaken and information collected in the OECM and around it.

However, the ability to readily access such public sites, and the completeness of the information contained there, are sufficient to establish this element of evidence. More detailed information would also need to be stored in national or regional (RFMOs) information systems.

5.5.4 Criterion D: Associated ecosystem functions and services and locally relevant values

5.5.4.1 Subcriterion D1: Ecosystem functions and services (EFSs)

Ecosystem functions and services are referred to in guiding principles (b), (c) and (f) as well as in criteria C3 and D1.

The elements of evidence suggested are: (i) EFSs *are supported, including those of importance to Indigenous Peoples and other local communities, for OECMs concerning their territories, taking into account interactions and trade-offs among ecosystem functions and services, with a view to ensuring positive biodiversity outcomes and equity (Evidence D1a); and (ii) management to enhance one particular ecosystem function or service does not impact negatively on the site's overall biological diversity (Evidence D1b).*

Given the intrinsic complexities of the concepts of ecosystem functions and services, and the wide range of factors they can include, assessing the degree to which this subcriterion is met can quickly become very demanding. Many functions might concur to produce one service, and one function may produce many services. Moreover, even though individual species have long been assumed to occupy unique niches (Chase and Leibold, 2003; Letten *et al.*, 2017), there is substantial functional redundancy in most marine systems (Rice *et al.*, 2013; Biggs *et al.*, 2020). In addition, more is constantly being learned about the extent to which EFs and ESs may be essential to humans (IPBES, 2019, 2024). In the case of cultural services, for example, support for communities' identity, it would be hard to define precisely all the functions that contribute to it. The same applies to all non-commercial values of OECMs.

General guidance on assessment and management of ecosystem services has been developing rapidly in the last two decades (e.g. MEA, 2005; UNEP, 2014[20]; Hattam *et al.*, 2015; IPBES, 2016; Salcone *et al.*, 2016), and the regional and global assessments of IPBES and follow-on thematic assessments (IPBES, 2018a,b,c) are beginning to inventory and map such information in a consistent manner. Nevertheless, information on specific areas is often scarce or absent, particularly in the marine realm, and uncertainty exists about the contribution of specific species or habitats to the processes, functions and services of the larger ecosystem.

For assessing this criterion, mapping the extent of the EFSs would be useful to assess the relative impact of the fishery and the effect of the OECM conservation measure. However, even such a mapping can be a complex, costly and uncertain task, far harder than determining the boundaries of a species or habitat distribution. Beyond mapping and describing the main EF and ES in the area, options for assessing how well the criterion is met become challenging. Not only is the task complex (as explained here), it also necessarily requires dealing with the substantial challenges of attaching 'value' to the various EFs and ESs (see Section 2.4.4). As a pragmatic approach, the initial assessment of the OECM could focus on getting descriptions of the major EFs and ESs that are as complete as available information allows, and highlight particularly EFs and ESs that would be expected to benefit from the positive and long-term biodiversity outcomes of the OECM. Then subsequent assessments could compare the unfolding EFs and ESs to these starting baselines, to evaluate effectiveness at delivering the function and services.

5.5.4.2 Subcriterion D2: Locally relevant values

Cultural, spiritual and locally relevant values are not mentioned in Aichi Target 11, but are part of the definition of protected areas (Dudley, 2008). They are mentioned in the definition of OECMs – underlining their importance in OECMs too – as well as in principle (J) and subcriteria C4 and D2. The two elements of evidence suggested (D2a and D2b) are examined below.

EVIDENCE D2A: CULTURAL, SPIRITUAL, SOCIOECONOMIC AND OTHER VALUES. The evidence should show that *governance and management measures identify, respect and uphold the cultural, spiritual, socioeconomic and other locally relevant values of the area, where such values exist.*

This requirement does not involve 'measuring' cultural, spiritual and other relevant values – which are in any case inherently impossible to quantify in individual cases (IPBES, 2022b). Instead, the need is to identify, respect and uphold those values, as they relate to the area and its biodiversity. What needs verification here is that a process was followed to accomplish that. The best way to do this is to involve in the assessment the people and cultures closest to or otherwise directly affected by the OECM, as effective participants (see Section 2.3.5.1). If those people and communities feel that as the

OECM is identified and used, their values have been 'identified', their contributions to the dialogue are respected (e.g. taken into account in decision making and assessments) and their values are upheld (e.g. politically and financially supported as appropriate), this evidence will be met. Subsequent assessments would return to the same communities to seek assurance that their values continue to be upheld and respected.

When conducting the OECM assessments, socioeconomic values can be readily identified and quantified, particularly for commercially directed fisheries, and should be available to assessments of OECMs. However, OECMs do not need to have high (or low) value for economic performance and provision of employment. The phrasing of this subobjective specifies that the OECM *respects* and *upholds* the various values. Consequently, for the economic and employment-related values, the initial OECM assessment should quantify to the extent possible the economic and employment performance, given the OECM measures in place or planning. Then the subsequent periodic assessments of progress can evaluate changes in those values. Changes in revenue and jobs alone are insufficient to judge if the OECM is upholding those values, because economic performance of a fishery can change for many reasons such as markets, competition from other seafood products, etc., and factors such as technological innovation can impact employment. However, if those are also tracked overall or outside the OECM, differential performance can give insight into how well those values are respected and upheld in the OECM.

EVIDENCE D2B: KNOWLEDGE, PRACTICES AND INSTITUTIONS. The evidence should show that *governance and management measures respect and uphold the knowledge, practices and institutions that are fundamental for the in situ conservation of biodiversity.*

The now outdated Aichi Target 18 already stated that *the traditional knowledge, innovations and practices of Indigenous and local communities relevant for the conservation and sustainable use of biodiversity, and their customary use of biological resources, are respected, subject to national legislation and relevant international obligations, and fully integrated and reflected in the implementation of the Convention with the full and effective participation of Indigenous and local communities, at all relevant levels.*

The need to consider local and Indigenous knowledge, institutions and practices in OECMs is repeatedly mentioned in various parts of Decision 14/8[21] as an enabling factor of effectiveness. Just as with the 'other values' referred to above, they relate to issues of 'participation', 'legitimacy' and 'equity' and are a fundamental element of 'equitable governance', buy-in and compliance.

The best way to ensure that the issue is properly addressed is to ensure an effective degree of participation in the OECM identification and implementation processes. Then the OECM assessment processes take the same approach as with cultural, spiritual and other relevant values (Evidence D2a). Only engaged communities can provide an evaluation of whether their knowledge and practices have been treated with adequate respect and subsequently upheld. The extensive guidance available on FPIC is useful when interacting with groups whose knowledge, practices and institutions are being considered (see references in Evidence B2b).

5.5.5   Considerations on additional properties

Decision 14/8 refers to important interrelated properties expected from OECMs, such as ecological representativeness, connectivity, complementarity and integration for which background information has been provided in Section 2.3.4. They are examined in this section in relation to their assessment in the identification and performance assessment processes. 'Integration' is considered in greater detail in Chapter 6.

These properties are mentioned sometimes in subcriteria (see Appendix 2), in the voluntary guidance or both. In this book, they are referred to as 'additional' because if strongly present, they support the OECM identification case, but if weak or absent do not disqualify an area from being an OECM when the criteria for identification have been adequately met. They are

nonetheless highly desirable as they would (i) enhance the OECM's effectiveness at conserving biodiversity in the site and in the broader ecological network; (ii) strengthen the rationale for identification; and (iii) point to possible future improvements of the OECM and conservation network effectiveness.

The additional properties are not simple to assess. They are mainly ecological in nature but depend also on institutional relations. They are continuous properties which may take a range of 'values' (and hence could be scaled) and not properties that may just be present or absent. However, their evaluation requires solid ecological information on the area of interest that may not be always available. Even with excellent collaboration between fisheries and environmental institutions, quantifying these additional properties in specific sites requires longer-term and expensive research. Moreover, there are no globally agreed standards of representativeness, connectivity or complementarity to which quantitative estimates of these properties could be compared. Qualitative expressions of these additional properties of OECMs may be developed by less demanding methods such as overlaying the fisheries and OECMs with ecological maps and using local ecological knowledge. Such qualitative and narrative information could be an adequate basis to have the assessment consider if the candidate OECM was particularly strong on the additional properties. Strengths or weaknesses of some of these additional properties could enhance or weaken the justification for OECM status, and their ecological importance should be taken into account when setting objectives for the OECM and developing management plans.

The development of MPA networks has often included assembling a large catalogue of regional biodiversity information to inform adoption of a biogeographic classification system for network design. This information, if available, should be examined before undertaking additional work on OECMs, underlining once more the importance of the collaboration of fisheries and conservation institutions, combining the science of area-based conservation and sustainable use. Strong involvement of the sector and the communities concerned cannot be overstated.

5.5.5.1 Ecological representativeness

Ecological representativeness is considered in Section 2.3.4.6a. Considering the ecological representativeness of the OCEM in a large biogeographic unit represents a significant broadening of the ecological relations between the fishery stocks and their associated and dependent species on the fishing ground. It requires more than local knowledge and personal fishers' experience and more connected institutional frameworks, establishing collaborations with environmental institutions. If a protected areas network already exists in the region, most of the information needed to examine this property in an OECM should already be available, facilitating the nesting of the OECM in the network.

From that perspective, the establishment of an effective network of MPAs complemented by OECMs would benefit from a systemic and iterative approach where known important areas (MPAs and OECMs) are described first and then remaining gaps in representation are identified and filled based on network criteria and management priorities (Johnson *et al.*, 2024).

Assessing an OECM's representativeness would require, first, finding an accepted biogeographic classification for the larger area in which the fishery and the potential OECM operate (e.g. the bioregion). On a global scale, the Global Ocean and Deep Seabed (GOODS) classification system (Vierros *et al.*, 2009) is widely used, but regional and subregional systems with finer resolution of ecological scales that have been developed in many areas can be used to facilitate the assessment. It would also require examining the position of the potential OECM in this spatial biogeographic classification. The judgement of representativity would take into account (i) the size of the potential OECM relative to what is known of the extension of the bioregion; (ii) the extent to which the boundaries of the potential OECM lie within the specific biogeographic category; (iii) the adequacy of its biodiversity values to represent such category or some mix of biodiversity from multiple biogeographical categories; (iv) whether that biogeographic category is underrepresented in the existing conservation network; (v) whether the OECM will increase 'representativity' at the network scale and at the scale of the biogeographic region.

A similar 'gap analysis' of the existing network relative to the full biogeographic region can also contribute to identify new areas (not yet identified as ABFMs) that could be identified as new OECMs and integrated into the FMP. Again, strong performance on each of those considerations is not essential, but each one, if documented, may increase the justification for considering the ABFM an OECM or suggest ways to improve the OECM performance.

5.5.5.2 Connectivity

As with the other additional properties, a practical assessment of connectivity might first focus on documenting the presence of any structural barriers to connectivity between area-based measures in the conservation network and between them and the rest of the ecosystem. These barriers would be a negative consideration in the assessment. This alone could be a major negative factor if some of the important biodiversity objectives were known to depend on connectivity among different ocean areas or oceanographic strata unless the structural barrier could be addressed in the management of the OECM.

In the earlier parts of the identification process, where functional connectivity was found to be important, evidence of barriers to the functional connectivity should already be sought, but often available information will be insufficient to firmly document either the presence or absence of barriers to functional connectivity. Furthermore, connectivity might be appreciated if the OECM area considered enhances the conservation network connectivity by filling eventual spatial or functional gaps and hence, presumably, improving ecosystem functioning and resilience.

Fully assessing the potential OECM's contribution to connectivity requires identifying the relations between the biodiversity attributes of concern in the potential OECM and the surrounding fishing ground and ecosystem, e.g. analysing the distribution and life cycle of key species, their migratory behaviour, continuities in bottom types and habitats, currents (for eggs and larval transport) and trophic relationships.

Consequently, finding information about such possible barriers could be an important consideration in designing the monitoring and successive assessments of the fishery-OECM, but the initial lack of certainty about degree of connectivity should not of itself impede OECM status if other criteria and subcriteria were satisfied adequately. Some of the necessary information would also have been used in assessing earlier criteria and subcriteria (e.g. C2 and D1), which would also have information about how the potential benefits provided by the OECM may enhance or augment the benefits provided by other OECMs and MPAs in the surrounding area, when relevant. However, as has been explained in those earlier sections, the important but varying connectivity of ecological processes is one of the major sources of uncertainty in assessing the actual biological and functional diversity of these systems, and how fisheries can affect them. In assessing the actual degree of connectivity for species, OECMs would have to merge all those other uncertain aspects, and would probably not be amenable to tight quantification.

5.5.5.3 Complementarity

The assessment of complementarity takes account of whether the OECM fills a gap in the biodiversity attributes protected in the ecological network, strengthens the functional connections among the network's areas or manages pressures or threats in ways that allow measures in other areas in the network to be more effective.

The concept indicates not only that the parameters used to measure biodiversity conservation outcomes of OECMs should be comparable to those used to measure the outcomes of MPAs, but also that OECMs and MPAs should be complementary in ensuring additionality and synergy. This implies that wherever MPAs exist in the vicinity of OECMs, connectivity channels are identified (e.g. life cycles, major connecting drivers, etc.). At the very least, OECMs should not provide biodiversity outcomes that would conflict with the objectives of MPAs to which they are functionally connected. Where an MPA could only provide partial protection to key biodiversity attributes (e.g. if the species migrates outside the MPA for part of its annual life cycle), additional protection provided by the neighbouring OECMs would be useful. Moreover, the fact that an MPA could be made more effective by

adding complementary measures in neighbouring areas could provide additional incentives for establishing OECMs in these areas. OECMs' complementarity may be realized in providing additional protected 'stepping-stone' areas in the life cycles of protected species, or in protecting critical habitats or food sources for these species. As appropriate, complementarity with other areas defined in the ocean for biodiversity-related purposes, such as ecologically or biologically significant marine areas (EBSAs) or key biodiversity areas (KBAs), might be considered.

For individual or sets of fishery-OECMs, that combined expertise should be able to evaluate and verify several important aspects of complementarity, including (i) adding biodiversity benefits either through its own direct biodiversity outcomes or enhancement of the effectiveness of other network areas; (ii) increasing the area coverage of the network; and (iii) improving or filling gaps in representativeness and connectivity. Presence of any of these factors is an additional positive consideration for OECM status but as 'additional considerations', their absence is not a serious shortcoming. However, at the very least, OECMs should not provide biodiversity outcomes that would conflict with the objectives of MPAs or other OECMs with which they interact functionally.

5.6 Step 4 – Synthesis and National Reporting

5.6.1 Synthesis of the full assessment

The task of the assessors in charge of the OECM identification is to evaluate the current and likely proximate future performance of an existing or new ABFM in relation to biodiversity conservation. After going through the first three steps, examining the extent to which a potential OECM meets each individual criterion, the assessors need to produce one synthetic conclusion for each potential OECM as to whether it could be considered a fishery-OECM or an upgradable ABFM (one that satisfactorily meets many but not all OECM requirements), or whether it remains solely a fisheries management measure. With so many criteria only possible to evaluate qualitatively or even by narrative, the synthesis will not be quantitative either. Individual criteria contribute to the synthesis in different ways.

- *Criterion A* (Section 5.5.1), on whether the potential OECM is already an MPA, has only one element of evidence, that emerges during the initial screening step. It is indeed a binary criterion for which the response can only be YES or NO: the area is an MPA or not. If the response is NO, the assessment can proceed further. If it is YES, the assessment has no need to proceed further. As an MPA, it is already part of the conservation networks, with outcomes dependent on how the MPA authority is managing the area. Whether the spatial fisheries measures are kept, adapted for additional outcomes or dropped altogether depends on how well the measures deliver the outcomes desired by the fishery management authority.
- *Criteria B–D* (sections 5.5.2–5.5.4) are more complex, requiring more elements of evidence with different properties. The response on each 'element of identification' will probably not be wholly 'positive or 'negative', and more likely will fall within a continuum between completely absent and completely present. Even considered separately for each element of evidence available for an individual subcriterion, there is likely to be both *some* identifiable or likely contribution to the desired positive biodiversity outcomes, and also some scope for further improvement.

When all the elements of evidence available for a given criterion have been examined, these qualitative judgements need to be combined to obtain an aggregate response for the criterion. Then the final aggregated assessment of the potential OECM performance requires combining the individual criterion assessments into a composite total one.

In some cases, major shortcomings in the potential OECM could be identified that could, until corrected, make the area unsuitable for OECM designation. Examples of these potentially fatal flaws may include the following.

- *Governance*: There are no agreements among legitimate international authorities in ABFMs located beyond or straddling across jurisdictions. There is a lack or insufficient inclusion of key stakeholders or rights-holders, including but not exclusively under the governance of domestic ABFMs by Indigenous Peoples and other local communities.
- *Size*: The area is much too large and institutionally complex to be credibly managed effectively. An example may be the fishery restricted area (FRA) adopted by GFCM, covering the entire Mediterranean sea bottom below 1000 meters depth (see Chapter 4). Conversely, the area is deemed too small to have, alone, any significant effect on biodiversity. This problem may be resolved by identifying a string of neighbouring and connected areas.
- *Fishery management*: The area is remote, not monitored, unenforced in practice or unlikely to ever be fished.
- *Biodiversity values*: Clear evidence of biodiversity values of concern, beyond target resources, is lacking.
- *Cross-sectoral issues*: Current or imminent activities of other sectors in the same area either remain a threat to the desired positive biodiversity outcomes or make long-term security unlikely.

Noting how much work is necessary in bringing together the multidisciplinary experts for the assessment (Section 5.2.5) and consolidating information (Section 5.3, Step 1), there could be benefit in completing the assessment of a failing ABFM, with the intent of giving integrated guidance on how to improve it to reach OECM status at some point in the future. That would be a positive return on the time and expense of the initial steps, and also continue to advance the mainstreaming of biodiversity in fisheries management and of sectoral management in conservation biology.

In cases where no 'fatal flaws' were found in the 'full assessment' phase and the subcriteria evaluations are sufficiently qualitative (at least placed in a relative gradient in some way), some sort of multiple criteria decision analysis (MCDA) might be feasible depending on the assessment context (see Fig. 5.3 for an example).

To proceed, the various conclusions of the assessment of different elements of evidence

Evidence		Sub-criteria		Criteria		Area
A	3	A	3	A	3	
B1a	3	B1	3			
B1b	3					
B2a	3	B2	2.5			
B2b	2			B	2.7	
B2c	2					
B2d	3					
B3a	2	B3	2.5			
B3b	3					
B3c	2					
B3d	3					
C1a	3	C1	2.8			2.9
C1b	3					
C1c	2					
C1d	3					
C2	3	C2	3	C	2.9	
C3	3	C3	3			
C4a	3	C4	2.8			
C4b	2					
C4c	3					
C4d	3					
D1a	2	D1	2.5			
D1b	3			D	2.8	
D2a	3	D2	3			
D2b	3					

Evidence	Colour Code	Score	Range	Decision
Good		3	2.6-3	OECM
Medium		2	1.6-2.5	To upgrade
Poor Absent		1	1-1.5	To drop

Fig. 5.3. (Left panel) Illustrative colour-coded representation of a potential OECM assessment synthesis. The conclusions of the assessments on each element of evidence (A to D2b) are colour coded and scored. Average scores are calculated for subcriteria (A to D2b), criteria (A to D) and the whole area. (Right panel) Example of decision rules. Criterion A score can only be 1 or 3. The red colour would be eliminatory.

and criteria must be standardized. Traffic light approaches have been used in fisheries, for example in Caddy (2002), Couture and Rideout (2014) and Gallacher *et al.* (2016). They require standardization of the evaluation of each element of information, in such a way as to lead to the same type of conclusion, e.g. 'yes', 'partially' and 'no' or 'good', 'medium' and 'poor'. These conclusions may be represented by a traffic light colour (e.g. green, yellow or red) or scored on an integer but not interval basis, (e.g. say from 1 to 3, without requiring the 'difference' from 1 to 2 being the same degree of difference as going from 2 to 3). Average scores may be calculated for each subcriterion, criteria

and for the whole area. At the end of the process the OECM assessment would be synthetized in a string of colours or scores corresponding to the string of elements of evidence (see Fig. 5.3 for illustration). A different type of scoring is shown in Table 5.1 and in Appendix 1.

The composite assessment against each criterion and the final assessment of the potential OECM performance need to combine the numerous individual assessments, whether colour-coded or scored, into a composite total one.

Decision 14/8 provides no guidance on how to reach the composite assessment. The decision rules needed (see Fig. 5.3, right panel) have such policy and operational implications

Table 5.1. Theoretical and simplified example of scoring of OECM properties in relation to governance and management (criterion B). Each 'element of evidence' (B1a to B3d) to be assessed for each subcriterion has been scored between 0 and 3 (Col. 2). The overall score for each subcriterion is given as a percentage of the maximum possible value for that subcriterion. The unweighted aggregated score for the whole criterion B is given in percentage at the bottom of the table. NR=not relevant in the area.

Properties	Scores	
(Options: 0=none; 1=poor; 2=medium; 3=good)	Number	%
B1: The area is geographically defined		
B1a: Size and area are known	2	
B1b: Boundaries are delineated	3	
B1 score	5/6	83%
B2: The area is governed		
B2a: To what extent is the legitimate governance identified?	3	
B2b: Have Indigenous Peoples been taken into account?	NR	NR
B2c: Is governance 'equitable' in CBD terms?	2	
B2d: Is governance collaborative enough to deal with threats?	2	
B2 score: OECM is eligible if B2 score >75%	7/9	77%
B3: The area is managed		
B3a: Area is managed to achieve biodiversity outcomes	2	
B3b: Relevant authorities are identified	3	
B3c: A management system is in place	3	
B3d: Management is consistent with the ecosystem approach	2	
B3 score: OECM is eligible if score >75%	10/12	83%
B: Unweighted aggregated score	22/27	84.1

that the legitimate authority may need to decide formally on this matter (instead of just the assessors), reflecting the degree of risk aversion desired in the policy and assessment. When areas under management by local communities can be considered for OECM status, even using a simple ranking rather than an integer scoring can be difficult to apply and scores awarded can be impossible to defend from critics not involved in the consensus-building process.

A standardized scoring process could reduce the burden for the legitimate authority and increase the overall coherence and equity of the assessment process across OECMs at national level and eventually at regional level (or across RFMOs). Any such scoring process would require decisions on factors such as the range of scores (e.g. 0–3 or 1–100%); the thresholds of categories within the range (e.g.0–30% = poor; 31–60% = medium; and >61% = good);

methods to aggregate scores across subcriteria and criteria (with weighting or not); and related decision rules (e.g.minimum score to be an OECM). If such standardization is desired, there would be advantages in efficiencies of process and consistency of outcomes if it were possible to fix and harmonize as much as possible the process and decision rules. However, any such standardization would have to take into account the potential differences in the amount and quality of information available and the level of threat and risk for the biodiversity and communities depending on it, and be acceptable at least in the fishery sector but preferably at national and if possible larger scales.

Until there is substantially more experience with evaluation of areas for OECM status, general recommendations on any specific approach to scorings are premature.

Figure 5.3 is only illustrative of the possible synthetizing process and, in principle, the assessment process may be decided case by case. For example, using a simple colour-coded approach, it might be decided that potential OECMs would be: (i) accepted as candidate-OECMs, to be submitted to the legitimate authority, if >80% of the elements of evidence are given a 'green' light; (ii) upgradable OECMs between 80 and 60%; and (iii) dropped below 60%. This, however, would give the same 'weight' to all elements of evidence which is illogical when some elements are necessary for OECM status whereas others are discretionary. This would also *de facto* make criteria and subcriteria with many elements carry more weight in scoring than ones with fewer elements. This may be addressed through giving different weights to different categories of evidence, to numbers of elements of evidence within a criterion, and across criteria. However, this then still imposes some other fixed weights on all the diverse types of evidence that may be encountered (see an example in Appendix 1). The illustrative example given above underlines the fact that, in most cases, the complex assessment of a fishery-OECM will not yield a simple YES or NO conclusion but one falling in a range within which a decision needs to be made.

As an example, an expert-based MCDA was used in the Aegean Sea by Petza *et al.* (2019) to assess over 516 broadly defined FRAs against seven OECM criteria[22] identified by experts with fisheries and environmental experiences (see Appendix 1). A MCDA using a decision tree has also been used within the FAO EAF framework, including in regional and national multistakeholder assessments in data-limited environments (Fletcher, 2008; Fletcher and Bianchi, 2014).

The same approach may be used to identify OECMs following the stepwise process described above to elaborate an aggregated score for potential OECMs (see Table 5.1).

When OECM status is approved for an ABFM that does need improvement in some elements of evidence or subcriteria, a risk is taken that the OECM might not improve as expected. Consequently, to mitigate that risk, the recurrent assessment required for OECMs should track the development, implementation and performance of such improvements and in cases of failure to improve, the OECM should simply be delisted.

If experience shows that too many upgradable OECMs fail to improve as expected despite the additional efforts, more stringent decision rules can be applied and fewer ABFMs will be granted OECM status. However, less risk-averse decision rules would be easier to implement and would provide incentives for the sector to improve its performance and promote both sustainable use and protection of biodiversity, consistent with the Convention itself and the preambles to both the Aichi Targets and the KM-GBF.

Regardless of the degree of risk aversion captured by the decision rules, there is weakness in the approach because they necessarily treat low scores on all subcriteria or elements of evidence as equally serious shortcomings in OECM performance, which is almost never going to be the case. That weakness can be reduced by weighting the subcriteria and elements by their 'importance', before scoring. However, those weightings are not going to be universal, because implications of shortcoming on any of the individual scores can depend on the ecological, economic or social context of the specific case. For these reasons, if a scoring approach was selected, the weighting of the criteria and subcriteria would rather be determined in a participatory process involving all actors.

Experience with application of the OECM criteria is at early stages, and is being acquired in widely separated marine areas and a wide range of ecological, economic, social and governance settings. Although coherence is highly desirable among assessments across a range of OECMs in a fishery, an EEZ or ecosystem, as is consistency of performance assessment over time (see Chapter 7), the expert community is still learning how to interpret and apply the criteria in the wide range of circumstances that will be encountered. To accelerate the building of coherence across applications, sharing evidence and experience will be important. The expert record of the assessment should include the information behind the scoring. This information should include: (i) the weights given to different types of evidence (whether formally or subjectively determined); (ii) the expert views on the value, strengths and weaknesses of the information available, including differences among experts when they arose and how they were resolved; and (iii) what factors were influential in the final decision on OECM status. Such information is likely to be valuable in subsequent phases of the OECM implementation cycle, for example, or when revisiting the decision to accept or reject the area as an OECM.

At the synthesis stage, these elements of evidence and their scoring rationales should be documented as fully as possible, in publicly available sites. If subsequent reassessment considered additional elements or types of information, or interpretations of previous evidence were expanded or altered, the rationales should also be documented for future reference.

In the short term, this documentation would be core information for setting or adapting the objectives, designing additional measures and updating management strategies and plans for the OECM. More widely, it would ensure transparency in the decisions made, and contribute to the growing body of information on how the criteria and elements are working. In the longer term, it will also be invaluable for future periodic reassessments, and eventual adjustments, if required.

Regarding the final advice or recommendation to be given to the legitimate authority after completing the scoring, Decision 14/8 does not specify any necessary minimum score, amount of improvement in biodiversity conservation[23] or level certainty about the outcomes that may be required to grant OECM status. This allows positive identifications to be made even when the OECM produces outcomes as diverse as, for example, (i) a strong positive outcome for one biodiversity feature of concern considered as extremely important (e.g. endangered whales, or turtles)[24] and (ii) moderate positive outcomes for a large range of biodiversity features of concern. This also allows positive identifications to be made for data-rich ABFMs managed by a well-resources central fishery jurisdiction as well as areas under robust local management with long culturally based knowledge of the area but few conventional monitoring data. The legitimate authority needs to have a good and faithful rationale for the decision. What is important is to clearly define the attributes of biodiversity claimed to benefit from the OECM, to provide evidence of such benefit, and of their likelihood to be maintained in the long term. Such evidence can only be the 'best evidence available', whether generated by complex surveys, powerful simulation models or local knowledge.

The Biodiversity Impact Mitigation (BIM) hierarchy promoted both by the IUCN and CBD may also be a helpful guide in building that rationale. In this framework, fishery-OECMs can be expected to contribute to: (i) avoid, when possible, impacts on the biodiversity attributes of concern within the OECM boundary; (ii) reduce/minimize unavoidable impacts; and (iii) mitigate the residual impact or facilitate recovery to reference benchmarks. The end result should be a reduction of the probability of occurrence of SAIs,[25] an increase in the security of healthy biodiversity attributes of concern and/or recovery of depleted ones, inside the fishery-OECM boundary and, for mobile elements, possibly also in the surrounding fishery and ecosystem.

At the end of the report produced by the assessors to inform the legitimate authority, the recommendations could be to consider the ABFM as one of the following.

- *Candidate OECM*, when the potential OECM satisfactorily meets the criteria. The candidate will be considered and endorsed (or otherwise) as a fishery-OECM by the legitimate authority, considering any other political and/or socioeconomic factors. If this positive recommendation is not upheld by the legitimate authority, documentation of the assessment should include the rationale for the decision.

- *Upgradable OECM*, when all the criteria are not satisfactorily met but opportunities exist for improvement of the weaknesses that were found. The report will include advice on possible improvement (e.g. modification of the boundaries and new technical measures). The legitimate authority would review all cost–benefit considerations and decide whether to pursue the improvements. If the authority agrees, the set of measures *intended to achieve* the expected outcomes should be specified and resourced. Then, if the necessary improvements are straightforward, a subset of the experts conducting the initial assessment might review them. If the desired outcomes are reasonably expected to be obtained in a short time, these experts could recommend that the ABFM be already identified as a fishery-OECM. Alternatively, the ABFM remains a potential OECM to be reconsidered at the next scheduled round of assessments or after a specified amount of time. If the legitimate authority does not considered it worthwhile, the upgradable ABFM will remain a simple ABFM.

- *Conventional ABFM*, when some criteria are not met and upgrades are not readily available. In this case, further consideration of the ABFM as a possible candidate OECM is not warranted. The area itself could be reconsidered for OECM status in the future but the management of the fisheries, or the other factors leading to the negative assessment result, would have to be quite different. In some cases, these ABFMs would have probably been detected during the quick screening step, but sometimes the more thorough assessment could detect shortcomings not revealed by the quick screening.

5.6.2 Reporting to the legitimate authority

We refer here to the report of the identification process to be submitted by the assessors to the legitimate authority and/or management authority at the end of the identification phase, for decision.[26] The final assessments (and possibly recommendations) elaborated by the assessors are presented to these authorities for final decision, in a coherent identification report, highlighting the rationale for the classification as (i) candidate OECM to be considered for formal recognition; (ii) upgradable ABFM worth considering for additional measures; or (iii) conventional ABFM to be kept in its usual fishery management role.

Legitimate authorities and managers often want, in addition to the full report, a concise, action-oriented report in the form of an executive summary or a more elaborated summary for policy makers,[27] for both internal and public uses. Such summary reports should always be accompanied by the comprehensive synthesis report described in the preceding section or provide a link to it.

The report should be comprehensive, highlighting the different recommendations but also their foreseeable consequences regarding,

for example, integration in the fisheries management plan, the recommended upgrading measures and what is known of their costs and benefits, and possible suggestions to improve the OECM network, etc. The assessors may also be asked for other types of recommendations that should be included and justified in the report. Because of the sectoral and potential cross-sectoral issues in some fishery-OECMs, the letter transmitting the assessment report to the legitimate authority might also recommend forwarding the report to other appropriate government or outside agencies, prior to its public release or posting, especially if those agencies contributed information and/or experts to the assessment process.

The report may conveniently be structured along the list of criteria followed in the identification process, including the different elements of evidence suggested in the Decision, and any additional information identified, used and important enough to be formally registered. If requested, the report will contain recommendations but otherwise, the presentation of the conclusions should be non-partisan, highlighting strengths and weaknesses relative to meeting the criteria, known costs and benefits, and major sources of uncertainties.

No single agreed structure or format for reporting the assessment and its results to the legitimate authority has yet emerged.[28] Many jurisdictions already have specified formats and document series for passing information from expert processes to decision makers, and these can be used whenever appropriate. They have advantages of being familiar to the decision makers and known to public interest who follow developments of the authority and areas under its jurisdiction, so these audiences are reached quickly. Whether part of such advisory series or developed for the particular case, the assessment report needs to include substantial information, although how details are apportioned among the core report, annexes to the report and supporting 'research reports' may differ among jurisdictional guidance.

Information that must be reported includes the following.

- An executive summary or summary for policy makers.

- The synthesis of the full assessment, summarizing briefly the OECM(s) performance on all criteria, and all subcriteria and elements of evidence that were influential in the synthesis.
- The conclusions (or recommendation) regarding OECM status, including a summary of the main justifications supporting the recommendation.
- Any other recommendations requested by the legitimate authority or arising from the assessment, with the respective main justification.
- A report on the area's performance on each criterion and subcriterion. The detail contained in the main report and in annexes or supporting documents may differ among jurisdictions, depending on their established practices, but should include, in the most suitable place:
  - a short description of all types of information contributed to the assessment
  - more detail, including figures, tables or expert narrative, on the subsets of information that were the basis for conclusions on how well or poorly the criteria were met[29]
  - brief justifications (often just a sentence) for why some contributed information was discounted or rejected by the assessors
  - if any formal scoring system was used,[30] the report would indicate: (i) whether the scoring system was provided by the legitimate authority, copied from use elsewhere (and the source) or developed during the assessment process; and (ii) the benchmarks or features that characterized each possible score. This may have to be done separately for each criterion or subcriterion if the benchmarks are not well-established properties such as maximum sustainable yield (MSY)
  - if no formal scoring system was used, the main considerations relevant to judging the criterion or subcriterion in this application
  - a clear statement (score or terse narrative) of how well the ABFM met the specific criterion or subcriterion, and all the substantive pieces of evidence supporting the conclusion for that criterion
  - any substantive pieces of evidence inconsistent with the conclusion for that criterion/subcriterion, and a short justification for why they did not determine the conclusion
  - any substantive uncertainties that arose during the assessment of that criterion/subcriterion and how they were handled in the conclusion
  - any differences among experts regarding the appropriate score/conclusion for the criterion, subcriterion or element of evidence, the competing justifications, and how the conclusion in the assessment was reached
  - other important components of the discussion during the assessment, such as possible follow-up research or management actions, implications for communities or perspectives not initially part of the assessment, etc.

It would also be useful to have a section of the report or an annex or supplement with general information about the area, even if not necessary for applying the criteria but informative regarding the area (e.g. its management and its ecological, economic and social contexts) and the assessment itself. For example: (i) chronology of past major management actions and ecological events in the area; (ii) dates of the assessment against OECM criteria[31]; (iii) oossible biogeographical subdivisions; (iv) all relevant levels and types of jurisdictions, including other economic sectors likely to be active in the area and potentially impacting its biodiversity; (v) main physical features like relief, depth range, bottom types; (vi) types of habitats like mudflats, mangroves, estuaries, lagoons, coral reefs, algal beds, seagrass beds, dynamic dunes, hot vents, seamounts, canyons, deep-sea corals or sponge reefs); (vii) hydrography parameters of relevance to the biodiversity functions, e.g. tides, currents, gyres, stratification (thermocline), turbulence; (viii) known natural threats such as severe storms, hurricanes, earthquakes, tsunamis and coral bleaching.

Some description of the broader area (ecosystem, ecoregion) within which the fishery and the OECM sit, e.g. based on available literature, would be useful. Many such elements would have been useful in the identification process, and could be downloaded in the WCMC database as 'additional information' for the sake of transparency.

If a comprehensive approach has been selected, assessing together a large sample of ABFMs, the report might also include conclusions at that broader scale (e.g. total coverage, representativeness, connectivity among OECMs and within conservation networks, etc.).

5.7 Step 5 – Decision by the Legitimate Authority

The legitimate authority receives the report of the identification process and has the prerogative to decide on the fate of the various potential OECMs, based on the information received and considering in addition any social, economic and political dimensions of the decision. Although Decision 14/8 encourages all parts of the OECM assessment and decision process to be effectively participative (see sections 2.3.5, 5.2.5 and 5.2.6), the actual decision is the responsibility of the legitimate authority, even when passed down to other governance processes. Participation of all main stakeholders includes ministries or other legitimate authorities with responsibility for conservation of biodiversity, and management of other sectors whose activities can have consequences for the desired positive biodiversity outcomes of the OECM. The decision should also consider its expected costs, benefits and other social and economic implications. Because of their dual objectives (sustainable use and biodiversity conservation) as well as possible cross-sectoral implications, a broader consultation and coordination may be felt necessary at this final stage of decision and, at the present time, there is neither experience nor guidance on circumstances when this would add value and when it would only increase the decision complexity and length of the processes.

If the elements to be considered under each step, their scoring system, their aggregation and the decision rules conditioning the decision options were formally agreed with the legitimate authority by relevant stakeholders at the beginning of the identification process, the decision could be fairly straightforward as only the conclusions need to be reviewed before decision. If this was not the case, however, the lack of transparency is likely to reduce buy-in and support and deter participation by the various groups identified in Decision 14/8. Moreover, many more aspects of the report will require detailed explanations and, in case of disagreements with some key elements of the assessment, like the decision rules used, may require reassessment. Notwithstanding, particularly given the little expertise available on fishery-OECM identification, the rationale behind these elements of the assessment might have to emerge from the assessment process itself, as proposals, to be formally endorsed together with the conclusions.

Following consideration of the report and possible recommendations, the legitimate authority must take decisions regarding (i) the identification of fishery-OECMs, a formal record of which should ideally be registered at national level; (ii) the potential updating of the FMP with its potential financial implications; (iii) the integration of the recognized OECM into the management plans of the fishery and/or the sector; (iv) whether to report to the WCMC for inclusion in the world OECM database[32] and consideration for reporting on international targets; and (v) when and how the information from the assessment will be made available to public access and use, including provisions for FPIC of the use of information from Indigenous Peoples, and protection of personal information on individuals, commercially confidential information and information treated as confidential by conservation authorities (e.g. breeding locations of legally protected species, etc.). The WCMC manual foresees that OECMs and information on OECMs might be reported by the legitimate authority or a range of data providers, preferably with the consent of the legitimate authorities. The actual status of OECMs reported by 'other providers' but without recognition by any legitimate authority has yet to be determined.

The granting of OECM status to an ABFM operating in a fishery may lead to the need to update the FMP, its objectives, measures, means, etc. and this updating may be fundamental for the integration of the OECM into the FMP (see Chapter 6) and a comprehensive appreciation of its performances (see Section 7.3). The updating needed may be relatively minor and within the mandate and budget of the management authority, in which case it is a formality. However, it may have more significant financial and operational implications, including in institutional and international collaborations, requiring a higher-level policy decision.

Finally, a policy decision might also be needed, and was probably made already at the onset of the OECM identification process at national level, regarding the policy decision to report all, some or none of the OECMs to the WCMC for future accounting against international targets. This issue is addressed more fully in Section 7.5.2.

Notes

[1] Including but not restricted to the fishery sector.

[2] As effectiveness has usually been assessed for the whole fishery, as a result of the package of spatial and non-spatial measures, and not for single measures.

[3] The Decision uses an implicit typology of outcomes related to their nature (e.g. biodiversity conservation, social economic, cultural, livelihoods); their direction (e.g. positive implicitly opposed to negative); their timing (actual/present/achieved, expected); their relation to policy and management (intended versus unintended), and the way they are identified and, implicitly, their degree of 'certainty' (e.g. actual, presumably observed and measured versus predicted by modelling based on various sources of knowledge and expert views).

[4] The Decision (Annex III, criterion B). Guidance for implementing the ecosystem approach under the CBD is found in Decisions V/6 (2000) and VII/11 (2004), and with respect to the ecosystem approach to fisheries, guidance is provided by FAO (2003a).

[5] CBD Decision 14/8 (2018), Annex III, criterion C.

[6] A management strategy evaluation is a computer simulation approach used to test different management strategies and options under different conditions and uncertainties, evaluating trade-offs and assessing the consequences of known uncertainties. It has also been used to evaluate the performance of marine spatial closures with conflicting fisheries and conservation objectives (Ditchmond et al., 2013).

[7] The comprehensive versus incremental assessment approaches are explained in section 5.2.1.

[8] ABFMs are formally established, spatially defined fishery management and/or conservation measures, implemented to achieve one or more intended fishery outcomes (CBD 2018, Annex IV, B, 2, c). When referring to ABFMs in this document, depending on context, we may refer to their spatial definition (the area) or to the specific management measures applying within them such as access rules, catch and effort limitations, gear specifications, and special by-catch regulations.

[9] There has been extended debate for decades on the possible vertical zoning of MPAs in marine multiple use environments in which different management may be needed at different depths to maintain existing fishing activities in the water column when the risk they represent for the biodiversity to be protected at the surface or on the bottom is minimal. (e.g. in Grober-Dunsmore et al., 2009; Davis, 2017). The recent IUCN-WCPA guidance in that respect for both MPAs and OECMs reflects a strong presumption against the use of vertical zoning of MPAs (Day et al., 2019) and OECMs (IUCN-WCPA, 2019) because of uncertainties about ecological bentho-pelagic coupling and enforcement difficulties.

[10] In the core text (paragraphs 4, 5, 6, 7, 10 and 11) as well as repeatedly in Annexes II, III and IV.

[11] Which states inter alia that each Party will... encourage the equitable sharing of the benefits arising from the utilization of such knowledge, innovations and practices.

[12] www.un.org/development/desa/indigenouspeoples/publications/2016/10/free-prior-and-informed-consent-an-indigenous-peoples-right-and-a-good-practice-for-local-communities-fao/ (accessed 23 July 2025).

[13] As available in the IUCN Red List of Threatened Species.

[14] EFSs are addressed in section 6.7, criterion D.

[15] In the case of mobile and highly migratory species, the distribution or life cycle of which cannot be circumscribed by the OECM.

[16] Some of the changes might be positive.

[17] NAFO/COM Doc. 23-28. Report of the NAFO Commission and its Subsidiary Bodies (STACTIC and STACFAD). 45th Annual Meeting of NAFO, 18–22 September 2023, Vigo, Spain. (Halifax: NAFO, 2023).

[18] In this regard, WG-EAFFM stated the following: '...at this time the working group recommends that the Commission request the Secretariat to submit VME sponge closures 1 to 6 and the seamount closures as OECMs to CBD in accordance with its procedures and to the United Nations Environment Programme World Conservation Monitoring Centre (UENP WCMC) for inclusion in the World Database on OECMs. The working group reflected that NAFO would need to update the CBD and the UNEP WCMC if the management measures are amended in the future.' NAFO/COM-SC Doc. 23-04. Report of the NAFO Joint Commission-Scientific Council Working Group on Ecosystem Approach Framework to Fisheries Management (WG-EAFFM) Meeting, 20–22 July 2023, Edinburgh, UK. (Halifax: NAFO, 2023), at 6.

[19] As an example, this concern is explicit in the UN Fish Stocks Agreement, for example, which aims 'to ensure the long-term conservation and sustainable use of straddling fish stocks and highly migratory fish stocks through effective implementation of the relevant provisions of the Convention' (Art. 2, emphasis added).

[20] See also www.gov.uk/guidance/ecosystems-services; https://oceanwealth.org/ecosystem-services/; https://roa.midatlanticocean.org/ocean-ecosystem-and-resources/characterizing-the-mid-atlantic-ocean-ecosystem/ecosystem-services/; and www.cbd.int/financial/monterreytradetech/unep-valuation-sids.pdf (accessed 25 July 2025).

[21] Principle (i): Annex II, Sections (A, 6), (B, 9), (B, 11, (i)); Annex III, Section (C, 1, f); Annex IV, Section (C, 1, b), (C, 3, e), (C, 4, c), (D, f, (ii))..

[22] CBD Decision 14/8 was not yet available at the time of the analysis and the criteria used were draft criteria emerging in the ongoing international discussions at CBD and IUCN.

[23] The benefit is only required to be obtained in situ, to be significant and positive, and to be maintained (or improved) in the long term.

[24] This targeted conservation approach is usual in protected areas considered as the more dedicated and effective conservation tools, if properly managed.

[25] The CBD Strategic Plan for Biodiversity 2011–2020 adopted the concept of significant adverse impact as a reference level for recovery plans and measures for threatened species and vulnerable ecosystems (in Target 6).

[26] Other reports may be sent by the legitimate authority to the WCMC for registration of the fishery-OECM in the world database, and also by the assessors on the performance of the fishery-OECM after integration in the fishery management plan (see Chapter 6).

[27] There is no standard format for a policy makers summary but good practice is to summarize the research and the key findings, the process used, the key debates and the points of consensus and controversy, and to clearly state the uncertainties or limitations of the assessment, and possible future improvements. The inclusion of central findings and figures that policy makers can quote in debates is essential.

[28] When When an OECM is registered with the WCMC, the registration material must have some specified properties (see section 7.5.2), but those become relevant only after the legitimate authority has decided on the OECM status for the area.

[29] Whether the detail is most efficiently presented early in the assessment report or presented under one of the criteria or subcriteria can vary, depending on whether the information was used in many criteria or a single one. If technical annexes or supporting technical documents are used, much of this detail can be in the technical support, although the main report should include important tables and figures that help readers understand the narrative in the report.

[30] If the same scoring system was used for multiple criteria or subcriteria, it only needs to be presented once, and referenced for its other uses in the assessment.

[31] The process may have covered days or months.

[32] Like the WDPA, the OECM database is a joint product of UNEP and IUCN, compiled and managed by UNEP-WCMC, in collaboration with governments, non-governmental organizations and other data providers. The database has been in development since 2019, in response to a request from parties to the CBD (2018).

6 Integration of Fishery-OECMs

Abstract

In order to be effective, a fishery-OECM needs to be functionally integrated in the management of: (i) the fishery within which it operates; (ii) other fisheries and economic sectors operating in the same area and impacting the biodiversity of concern to be protected; (iii) and the conservation network of the seascape and ecosystem to which it belongs. This chapter considers the practical integration of fishery-OECMs at these different scales, emphasizing the importance of aligning the management measures to ensure coherent and effective conservation outcomes. It discusses the need for co-ordination to avoid conflicts, enhance synergies and address the challenges and opportunities of integrating fishery-OECMs across multiple uses and jurisdictions.

In order to be effectively managed, and to benefit from the fishery management system capacity, a fishery-OECM needs to be functionally integrated in the management plan of the fishery within which it operates. Ideally, to reliably be able to deliver the desired broad positive biodiversity outcomes, the fishery-OECM management should also be functionally integrated with the management of other fisheries, fishery-OECMs and other economic sectors[1] operating in the adjacent surroundings, if their activities could affect the desired or expected biodiversity outcomes. Fishery-OECMs should also be integrated within conservation networks, together with MPAs, developing synergies and complementarities at EEZ, regional, seascape or ecosystem level.[2] These different levels of integration are addressed briefly in the following sections, focusing on the fishery sector.

The different institutional and spatial scales of integration call for increasingly complex governance arrangements and levels of capacity. The integration of a fishery-OECM into the fishery management plan, and in the fishery sector, are addressed in the next two sections. The broader levels of integration, across economic sectors and within conservation networks, are beyond the strict management mandate of fisheries authorities, and require not only technical but also institutional collaboration frameworks. In addition, regardless of the scale of integration, the information and capacity developed to integrate fishery-OECMs within the fishery sector will be useful for the broader integration levels that the governance processes must address.

6.1 Integration of OECMs within the Fishery Management Plan

Abundant guidance is available for the management of responsible fisheries, in the Code of Conduct for Responsible Fisheries (FAO, 1995) and the related guidelines on management, on the precautionary approach and on the ecosystem approach to fisheries (EAF) (FAO, 1996a, b, c, 2000, 2003, 2009a, b, c, 2011, 2015; Cochrane and Garcia, 2009). Consequently, the following sections focus on the management issues specifically related to fishery-OECMs and their integration in the fishery management plan (FMP).

Usually, existing ABFMs – with the technical measures applied in their area – are already integrated in the FMP (or equivalent for locally managed areas) for enforcement, monitoring and evaluation. In principle, their effectiveness in contributing positive outcomes for the fishery justifies their existence, even though such effectiveness is rarely recurrently assessed. However, when given an OECM status, the ABFM is expected to generate additional positive outcomes for biodiversity features of concern for which new objectives, measures and implementation means may be needed in the FMP.

The enabling frameworks of relevance to OECMs' management effectiveness were reviewed in Chapter 4. The enabling factors include: (i) the standing of the legitimate authority; (ii) a formal fishery management plan or equivalent traditional set of rules; (iii) equitable governance, as specified in Decision 14/8; (iv) enabling international instruments particularly when resources and biodiversity attributes of concern are transboundary, straddling or exclusively in the high sea; and (v) adequate management capacity including deterrent enforcement. These factors apply *mutatis mutandis* to fishery-OECMs.

Additional factors may have to be considered in fishery-OECMs such as: (i) knowledge of current or likely threats from other economic sectors or natural drivers of system dynamics and their impacts; (ii) identification or foresight of climate change impacts on biodiversity and related responses, different from fisheries impact but interacting with it; (iii) enhanced knowledge on the biodiversity attributes of concern and expected benefits of the OECM not previously considered; (iv) the pressures exerted in the past, currently and in the future; and (v) the likely resulting threats on biodiversity, as well as the potential benefits realized (or expected from) the OECM. In addition, the introduction or reinforcement of a recurrent monitoring, evaluation and reporting (MER) programme is addressed in detail in Chapter 7.

The formal integration of the fishery-OECM(s) into the FMP aims to (i) take advantage of the resources already invested in management of the fishery and the ABFM function to manage the OECM functions; and (ii) increase coherence between the fishing and conservation regimes implemented inside and outside the OECM, contributing to improving effectiveness in relation to objectives and efficiency within a usually very tight budget. It is likely that the means available to monitor and assess an OECM will be commensurate with (and could benefit from) the resources available for monitoring the whole fishery, making integration desirable not just in the application of the conservation measures, but in monitoring and assessment of the fishery outcomes and biodiversity outcomes. In small-scale fisheries, particularly in community-managed areas, the means available for

management, monitoring and assessment may be limited, but the spirit of integration ought to be apparent. If the biodiversity attributes of concern targeted by the OECM were already the subject of conservation activities of other institutions (like seabirds, marine mammals or marine turtles), coherence between the respective management plans, monitoring programmes and assessments would be both efficient and cost-effective, and reduce the likelihood of later disagreements among the fishery and the biodiversity experts over effectiveness in delivering the biodiversity outcomes.

The same rationale applies to the upgradable ABFMs – those for which all the criteria are not satisfactorily met but where opportunities exist for improvement.

The integration of fishery-OECMs and upgradable ABFMs may require::

6.1.1 Noting formally the OECM(s) and upgradable ABFMs to be covered by the FMP in the scope of the management planning document

The descriptions of the OECM(s) and upgradable ABFMs in terms of the area's characteristics and the specific conservation measures applying inside them may usefully be annexed to the FMP, with their specific characteristics (see below).

6.1.2 Updating the FMP objectives and targets to better reflect the specific biodiversity conservation objectives and expected outcomes of the fishery-OECM(s)

The nature of the objectives and targets regarding the desired biodiversity outcomes that could be added are presented in Section 5.5.3.2, criterion C1a.

6.1.3 Specifying the indicators and reference values or trends

Specifying the indicators and reference values or trends and other performance benchmarks or standards related to the above objectives, and directly connected to the identification criteria, that are needed for the future recurrent assessment of the OECM performance. The MER sampling and assessment programmes are adjusted accordingly (cf. Chapter 7).

6.1.4 Specifying the measures taken in the OECMs to reach the objectives

The individual measures might or might not be area based, and their objectives could be to reduce impact on non-target species, protected species and vulnerable habitats affected by significant adverse impacts (SAIs; see section 5.5.3.2 for details); maintain a functional ecosystem structure (see section 5.5.4) or address the 'additional properties' (see section 5.5.5). Some measures intended to achieve specific OECM objectives could even provide incentives to enhance the ability or willingness of the fishery actors to deliver the desired biodiversity outcomes, even when no specific biodiversity objectives have been mentioned when the measures were included in the FMP.

Special measures might modify: (i) existing rules of access to the area; (ii) fishing gear specifications; (iii) catch and by-catch regulations (particularly on threatened species); (iv) habitat protections and restoration measures; (v) logbooks and on-board observer manuals; (vi) electronic navigation and vessel monitoring systems (VMS). These modifications should also improve the coherence (complementarity and synergy) between the measures applied in the fishery-OECM and around it. For traditional fisheries lacking a formal FMP, the prevailing local management rules may need to be upgraded (if needed) by the local legitimate authority. The recurrent reporting to the CBD and WCMC would imply that although informal, these measures should be somewhat registered at least at the level of the legitimate authority, with any possible evidence that they are effectively applied.

6.1.5 Strengthening participation

Strengthening participation of environmental, civil society and community stakeholders,

in addition to fisheries stakeholders, in the development of the FMP. Active stakeholders' participation in fishery management is already an established best practice and thus presumably also in the management of the ABFM. The fishery-OECM process will have to follow this practice (see section 5.5.2.4). If this was not yet the case, the implementation of an OECM presents the necessity to establish or strengthen participation, considering that fishery stakeholders as well as environmental, civil society and community interests are a source of knowledge on the environment and biodiversity. In particular, if the location of the OECMs is likely to impact a *specific* coastal community or stakeholder group (e.g. environmental and sectoral NGOs), their participation is advised under criterion B2b, and would improve effectiveness and equity (see below).

6.1.6 Broadening the communication target audience

Broadening the communication target audience to inform interested parties about the presence of OECMs in the fishery and their implications for the fishery itself; for other fisheries operating in the same ecosystem; as far as possible, for other economic sectors that might help promote the OECM (e.g. ecotourism) or might have to revise their policies or practices to allow the desired OECM biodiversity outcomes to be fully realized. The objectives are to promote a good understanding of the new measures; increase collaboration; inform all actors of the consequences of non-compliance; and hence, hopefully, improve OECM performance.

6.1.7 Checking and dealing with additional equity issues

Checking and dealing with additional equity issues potentially created by the OECM status beyond those already addressed in the old ABFM – especially in the light of cultural, spiritual, socioeconomic and other locally relevant values, as discussed earlier. Issues that might need to be dealt with could include, for example, (i) additional disruption of traditional livelihoods, including relocations of fisheries from traditional areas; (ii) new or increased violations of cultural or spiritual values; and (iii) changes in sharing or access arrangements among fishery participants that alter distribution of costs or benefits, including both among geographically dispersed fisheries and within fisheries with individual quota allocations; or (iv) participants that have to exit from the sector because they cannot comply with all the necessary spatial or non-spatial measures of the OECM.

New measures may need to be introduced to address new equity issues by, for example, (i) adapting fishery measures in and out of the OECM to mitigate the distortions; and (ii) introducing alternative measures such as additional income-generating activities (AIGAS) or other forms of compensation, aimed at addressing inequities while preserving the fishery-OECM status and conservation objectives. This would be particularly important when both small-scale and large-scale fisheries are involved.

Evaluating risk of non-compliance with OECM measures and strengthening MCS around and in the OECM. Where the risk of non-compliance justifies it, identify impeding factors and corrective measures taken to mitigate their effect, as well as opportunities to incentivize improved compliance.

6.1.8 Addressing impending internal and external threats to OECMs

Addressing impending internal and external threats to OECMs and clarifying contingency measures and monitoring activities and benchmarks, in order to detect and respond quickly to emerging threats, improve resilience of the fishery and the OECM, and optimize the long-term total biodiversity benefits and co-benefits. This activity requires fisheries managers to, *inter alia*: (i) identify the elements potentially at risk in the natural (ecosystem components, functions and services) and human components (employment, livelihoods, cultural practices) of the fishery system, and the sources of threats, in the environment, the fishery or in other economic activities; (ii) improve foresight and predictive capacity: developing the needed threat-specific competences and collaborations with the

relevant experts, and collecting the relevant information through the MER system; (iii) for risks considered large in (i), assess as fully as feasible the likelihood and possible magnitude of the related risks[3]; (iv) identify responses to threats that are robust to uncertainty[4]; (v) develop contingency plans and associated triggers for action; (vi) display transparent information on uncertainties and their potential consequences for decision making and implementation; (vii) develop a schedule to periodically review the threat identification and assessments, and update measures and contingency plans; and (viii) when relevant, identify/strengthen the regional collaborations needed to address transboundary threats.

6.1.9 Ensuring that the fishery management plan is adaptive

That (i) it has considered potential changes that might occur in the OECM and/or in the fishery system which would affect the OECM's performance; and (ii) it includes procedures for their early detection and when necessary, to maintain effectiveness in delivering fishery or biodiversity outcomes, their mitigation. Effective decision rules avoid over-responding to small oscillations in performance due to natural variability in the social-ecological system. That feature is particularly important when pursuing longer term positive biodiversity outcomes, when the biodiversity features may be vulnerable to different environmental pressures than the traditional target species of the fishery. When information allows, best practices include defining thresholds (e.g. in preagreed decision rules) of change in indicators beyond which management responses would be justified. The central role of the MER system in this regard is developed in Chapter 7.

6.1.10 Archiving and maintain information on FMP provisions and implementation

This function is best undertaken by a well-equipped MER system (see Chapter 7).

It should be reiterated that, as for the identification, the integration of OECMs in FMPs in specific cases may be affected by differences in resources, capacities and governance constraints, but in each case the integration needs to be explicit and credible.

6.2 Integrating OECMs within the Broader Fishery Sector

As one of the intents of OECMs is to contribute to biodiversity conservation at the ecosystem level, it would be desirable, subject to resources and capacity available, to co-ordinate and harmonize the identification, management and performance assessment of OECMs among fisheries that exploit the same ecosystem and food chain, contributing to integration of biodiversity concerns and measures into the sector.

Decision 14/8 (§12) urges Parties to *facilitate mainstreaming of ... other effective area-based conservation measures into key sectors, such as,* inter alia, *...fisheries.* Mainstreaming and integration of biodiversity concerns into the fishery sector were addressed specifically in 2016 in Decision 13/3, which: (i) in paragraph 62, encourages fisheries management organizations to further consider biodiversity-related matters in fisheries management in line with the ecosystem approach, including through interagency collaboration and with the full and meaningful participation of Indigenous Peoples and other local communities; (ii) in paragraph 63, re-emphasizes the importance of collaborating with FAO, RFMO/As and RSCs when addressing biodiversity conservation in sustainable fisheries; (iii) in paragraph 68, urges parties to use existing guidance related to EAF and calls for further collaboration and information sharing among the CBD Secretariat, the FAO and regional fishery bodies regarding information on EBSAs and VMEs, in support of achieving Aichi Targets.[5]

The "integration" of the fishery-specific OECMs across the entire fishery sector would be institutionally simpler and, in some cases, less expensive in additional interaction costs than integration with other relevant economic sectors, or within MPA networks (see sections 6.3 and 6.4), as it could be undertaken within one line ministry and legal framework, using many of the same collaborations with environmental agencies that are already needed for the fishery-OECMs. It would also give to EAF a further boost away from the myopic and criticized single-species, single-fishery approach.

Despite rising awareness of the systemic nature of fisheries in the last 50 years, the adoption of the ecosystem approach to fisheries (EAF) at FAO in 2001 (FAO, 2003; Garcia *et al.*, 2003) and progress made in the effective integration of single fisheries' management at whole-sector (or ecosystem) level (e.g. Walters and Hilborn, 1976; Walters, 1980; Allen and McGlade, 1987; Charles, 2001; Garcia and Charles, 2007; Link, 2018; Link *et al.*, 2020), substantial scope for further integration still exists. Integration taking account of fisheries impact on the trophic web has been explored from a theoretical scientific angle, focusing on issues and approaches such as multispecies maximum sustainable yield (MSY), ecosystem-wide MSY, system-level optimum yield and the portfolio approach (Link, 2018), 'regional resources' (Fletcher *et al.*, 2010, 2012) and balanced harvest (Garcia *et al.*, 2012; Zhou *et al.*, 2019).

In operational management, this has led to considering multiple fisheries on several target-species or assemblages (like most small-scale fisheries) as "single" multispecies-multigear fisheries, albeit with little real change in management. This has led also to innovative management approaches such as: (i) implementing an overall cap on total ecosystem catch in the eastern Bering Sea and Gulf of Alaska LMEs (Link, personal communication); (ii) limiting removals from prey (forage) species stocks in the Antarctic (Constable, 2011); or (iii) considering together for management the various fisheries (métiers) using a common species assemblage (referred to as a "resource") as in Western Australia (Fletcher *et al.*, 2010, 2012). Other factors promoting integration of management across fisheries include considerations such as allocating fishery-specific harvest caps on by-catch of protected, endangered or threatened (PET) species[6] and co-ordinating supply to the processing sector, to promote stable employment and maximize value of the catch.

In many countries, the management of large-scale fisheries tends to follow a "western" quantitative, reductionist and science-driven approach, including often detailed monitoring, while the management of small-scale multigear-multispecies fisheries rests on a deeper understanding of multiple and detailed sources of knowledge on resources and sociocultural aspects, with little or no continuous monitoring. Integrating both types of fisheries may be a challenge

In facing the integration challenges, there must be a strong recognition that the extent to which a particular OECM can engage in a massive integration across multiple fisheries (or across sectors) will depend on resources and capacity available. For example, a longstanding local-level community-managed area can be producing strong biodiversity benefits in a situation of minimal resources, and trying to force it into "integration" may do more harm than good – and would require an influx of large levels of resources and capacity.

Moreover, the "integration" of the bio-ecological and socioeconomic dimensions of different fisheries in an ecosystem could go to a variety of depths, from information exchange to effective co-ordination of selected measures among sets of related fisheries, and to full integration of the management plans. The workload of initial integration grows as the number of fisheries increases, but the eventual costs and workload of the integrated management of multiple fisheries might provide efficiencies in both MER and marketing that more than offset the initial costs and effort.

Subject to resources and capacity available, activities towards integration of fishery-OECMs at the fishery sector level would include: (i) Mapping all fisheries footprints (spatial distribution of fishing effort) and OECMs.; (ii) Looking for potential synergies among the OECMs used in various fisheries in terms of geographical and functional connectivity; (iii) looking for good "marriages" of common or complementary measures (spatial and non-spatial) that could improve integration of fisheries management among fisheries using OECMs or not, and facilitate delivery of the biodiversity outcomes and (iv) Harmonizing or, where feasible, merging management plans and measures of strongly overlapping or complementary OECMs and other FMPs, to facilitate their management (e.g., through economies of scale in monitoring and enforcement) and, possibly, optimize their biodiversity outcomes. The merging of

OECMs, when functionally appropriate, may be an effective way to enhance their performance.

6.3 Mainstreaming OECMs across Economic Sectors

Mainstreaming biodiversity is:

> The general process of embedding biodiversity considerations into policies, strategies and practices of key public and private actors that impact or rely on biodiversity, so that it is conserved and sustainably and equitably used. (Huntley and Redford, 2014)

A more specific definition suggested specifically for fisheries is:

> The progressive, interactive process of recognizing the values of biodiverse natural systems in the development and management of fisheries, accepting full accountability for, and effectively responding to, the broader impact of fishing and fishery related activities on biodiversity and related structure and function of ecosystems. (Friedman et al., 2018)

Mainstreaming OECMs across sectors can be seen as a way of mainstreaming biodiversity concerns through the use of area-based conservation tools. Paraphrasing the above definitions, mainstreaming OECMs across sectors could be defined as:

> Embedding OECMs into sectoral development policies and management strategies and practices, addressing the sectoral impact on biodiversity and ecosystem services, so that they are conserved and sustainably and equitably used.

In Decision 14/8, the terms "mainstreaming" and "integration" are used interchangeably.

In 2018, FAO adopted a strategy for mainstreaming biodiversity at national, regional and international levels, in the sectors under its mandate: agriculture (including livestock and crops), forestry and fisheries (including capture fisheries and aquaculture). The strategy aims to (i) promote sustainable use and management of biodiversity; (ii) conserve, enhance and restore biodiversity, and ensure provision of ecosystem services; (iii) integrate conservation, recognition and promotion of biodiversity throughout the value chains;

and (iv) safeguard livelihoods of small-scale producers and Indigenous Peoples and local communities. It relies *inter alia* on effective governance, partnerships, knowledge-based approaches, inclusive governance and equity. Planned activities include: (i) capacity building at national level; (ii) mainstreaming of biodiversity across FAO programmes, policies and activities; and (iii) advocacy for recognition of the role of biodiversity for food security and nutrition (FAO, 2019b, 2021). These efforts extend those made for decades, at least since the United Nations Conference on Sustainable Development (Rio de Janeiro, 1992) with the same aim, albeit using sometimes a different language and with mixed success.

Decision 14/8 (Annex I, Section II, B) provides guidance about the integration of OECMs across economic sectors. Paragraph 2 of the Annex indicates that such integration could be achieved *by applying the ecosystem approach and taking into account ecological connectivity and the concept, where appropriate, of ecological networks.* Exploring all the necessary features and challenges of cross-sectoral levels of integration is beyond the scope of this book, but the considerations made above (in sections 6.1 and 6.2) for single-fisheries and integration across the fishery sector all apply to the participation of the fishery sector to cross-sectoral co-ordination frameworks, when established at the appropriate level of the government. This type of integration is more demanding and would usually require some overarching framework such as ICAM (United Nations, 1992a), Integrated Coastal and Ocean Management (ICOM; Belfiore *et al.*, 2004),[7] Integrated Ocean Management (IOM; Freestone *et al.*, 2010)[8] and MSP (Jentoft, 2017; Wright *et al.*, 2018). Such structured approaches have been advocated for several decades, often meeting practical implementation problems and producing mixed results.

In coastal areas, as in highly populated land areas, potential OECMs may often occur in areas in which multiple uses are allowed under various authorities, complicating the establishment of 'pure' fishery-OECMs. In the open ocean pelagic domain, the likelihood is that fisheries and navigation often might be the main if not only interacting sectors. This fact gives to RFMOs operating over deep oceans an important opportunity to identify fishery-OECMs as the areas

under their mandate is large, covering many areas in which only or mainly fisheries operate.

Based on in Decision 14/8 (Annex 1, II, B), the actions required for such integration could be: (i) To Identify, map and prioritize areas important for biodiversity attributes of concern and essential ecosystem functions and services; (ii) To map both the spatial footprints of the different economic sectors on the biodiversity features in (i) and of the spatial measures used in each sector; (iii) To consider merging strongly overlapping sectoral OECMs for practical operational purposes to the extent feasible; (iv) To look for effective combinations of common or complementary measures across sectors, that could facilitate delivery of the biodiversity outcomes; (v) To harmonize the relevant sectoral legislations as necessary to enhance complementarity; (vi) To review and update sectoral plans as necessary to ensure that they both recognize and incorporate the many values provided by OECMs, and do not impede delivery of the positive biodiversity outcomes of any sectoral OECMs; (vii) To develop targeted communications campaigns that lay out the biodiversity and ecosystem functions and services provided by OECMs; (viii) To review and revisit existing policy and finance frameworks to improve the enabling policy and financial environment for sectoral mainstreaming; and (ix) To assess and update the capacities required to improve the synergetic mainstreaming of protected areas and OECMs, for example improving enabling policy environments and improving coordination among sectoral systems of monitoring and evaluation.

6.4 Integrating OECMs in Seascapes

Decision 14/8 (Annex I) also addresses the integration of OECMs in broader governance areas like landscapes and seascapes. Conservation International defines seascapes as

> Large, multiple-use marine areas, defined scientifically and strategically, in which government authorities, private organizations, and other stakeholders co-operate to conserve the diversity and abundance of marine life and

to promote human well-being (Atkinson *et al.*, 2011:2).

IUCN defines the MPAs in Category V 'seascapes', i.e:

> areas where the interaction of people and nature over time has produced ... distinct characters with significant ecological, biological, cultural and scenic value, and where safeguarding the integrity of this interaction is vital to protecting and sustaining the area and its associated nature conservation and other values.[9] (see also Dudley, 2008; Day *et al.*, 2019).

The two definitions underline that fact that, in the context of biodiversity conservation, (i) seascapes are governance framework created by humans; and (ii) they are arenas of active interaction between humans and nature for broad-scale conservation of biodiversity and ecosystem services and other societal values, recognizing traditional management practices and providing models of sustainability.[20] The 'seascape' approach is aimed at building coalitions among government(s), corporations and civil society to improve ocean governance and highlight the importance of achieving effective governance across sectors and at all levels, from local to regional (linking sections 6.3 and 6.4). It calls for bringing in the necessary science and knowledge, and empowering local governments and communities. Seascapes may be national, such as the Bird's Head seascape in Indonesia or the Abrolhos seascape in Brazil. They can also be international, such as the Eastern Tropical Pacific Seascape of Costa Rica, Panama, Colombia and Ecuador, or the Sulu-Sulawesi Seascape of The Philippines, Malaysia, Indonesia, Papua New Guinea, the Solomon Islands and Timor Leste. Guidance on seascape establishment, governance and management is available, for example in Ervin *et al.* (2010) and Atkinson *et al.* (2011).

This level of integration of OECMs in seascapes is beyond the management mandate of any single fisheries authority, whether national or regional, and extends beyond the scope of this book. However, the elements of guidance provided in sections 6.2 and 6.3 and the preparations for integration of OECMs across the whole fishery sector and other economic sectors in specific areas would assist in such broader engagement with other sectors on seascape scales.

The integration of OECMs in seascapes is explicitly addressed in Decision 14/8 (Annex 1, A) as a strategy to combat ecosystem fragmentation and to optimize the functional performance of individual MPAs and OECMs through improved connectivity. The suggested actions in this Annex are: (i) To review national visions, goals, and targets to ensure that they include elements of integration of protected areas and OECMs; (ii) To identify key species, ecosystems, and ecological processes, including those vulnerable to climate change; (iii) To identify and prioritize important areas (including OECMs and MPAs) to improve connectivity and mitigate the impacts of fragmentation; (iv) To conduct a national review of the status and trends of habitat fragmentation and connectivity for key species, ecosystems and ecological processes, and to review the role of MPAs and in connectivity; (v) To identify and prioritize actions with the sectors responsible for habitat fragmentation, and engage them in mitigating their impacts on conservation networks; (vi) To review and adapt seascape plans including marine spatial plans, sectoral plans, or integrated marine and coastal area management plans; and (vii) To prioritize and implement measures to decrease habitat fragmentation and increase connectivity, including PAs, OECMs, and indigenous and community conserved areas (ICCAs).

Notes

[1] Decision 14/8, paragraphs 1 and 4 and in Annexes I, III and IV.
[2] Cf. Decision 14/8, paragraph 7; extensively in Annex II on governance and management; in Annex III in the guiding principles and criteria for identification; and in Annex IV on achieving Target 11.
[3] Conventionally measured by the potential cost of a damage multiplied by the probability that such damage really occurs.

[4] For example, using mmanagement strategy evaluation (MSE) – if the capacity and necessary information are available or accessible – or developing risk-based decision rules based on available information including expert opinion and local knowledge.

[5] https://www.cbd.int/doc/decisions/cop-13/cop-13-dec-03-en.pdf.

[6] For example, in the mixed species rockfish fishery in British Columbia, the mixed species flatfish fishery on the Scotian shelf and in MSC certifications requiring to keeping by-catches under specifed limits.

[7] See also: https://globaloceanforum.com/areas-of-focus/integrated-ocean-and-coastal-management/

[8] See also: (1) http://www.beaufortseapartnership.ca/integrated-ocean-management/integrated-oceans-management-plan/; (2) https://www.oceanpanel.org/blue-papers/integrated-ocean-management

[9] See also: (1) https://www.biodiversitya-z.org/content/iucn-category-v-protected-landscape-seascape; (2) https://www.conservation.org/priorities/seascapes-large-scale-marine-management

7 Monitoring, Performance Evaluation and Reporting

Abstract

This chapter stresses the importance, for a fishery-OECM, of recurrent evaluations to demonstrate the continuous production of the expected biodiversity benefits that justify its status. The monitoring, evaluation and reporting (MER) of the OECM performance are functionally and operationally integrated into those of the broader fishery-MER. The chapter indicates the elements that need to be specifically considered in strategic planning of the OECM-MER to assess: (i) the performance of the OECM in protecting biodiversity, and (ii) the efficiency of the MER activities in providing the information needed. 'Effectiveness' is examined in detail, including definition, requirements, objectives, complexity, scope, methodology, multidisciplinarity, factors of performance and likelihood of fishery-OECMs performance. The process and requirements for reporting on performance, at national level and to the CBD and UNEP-WCMC, are examined.

7.1 Preliminary Considerations

For their effective adaptive management, all fisheries ought to be monitored to systematically collect data and generate the scientific and traditional knowledge regarding the fisheries, the natural resources they use, the status of oceanographic and ecosystem drivers that could affect those natural resources, the impact of the fisheries on target species, their compliance with management measures and the outcomes of the actual management of the fishery. This requires at least periodic overviews of the status of the resources, the fisheries, the social-ecological system and the effectiveness of the management arrangements and measures. That is the role of monitoring, evaluation and reporting (MER) activities, undertaken by more or less sophisticated settings, depending on the national or local governance capacity, the fisheries' size, and economic and social values of biodiversity components impacted by the fisheries. Indeed, even the term MER may seem onerous in some informal contexts.

Most if not all fisheries management strategies use ABFMs as a spatial management measure within which fishing practices are controlled to maintain target, associated and dependent species, essential habitats and/or vulnerable life stages, as well as to reduce by-catch of unwanted or protected species. After being introduced, ABFMs may be evaluated on an *AD hoc* basis, e.g. if the fishery fails to operate as expected.[1] The fishery MER usually assesses only the overall performance of the management of the entire fishery, looking at trends in key fisheries elements like fishing mortality, abundance of target species, recruitment, spawning biomass, by-catch of unwanted or protected species, sometimes state of essential habitats, etc. That said, the focus and rigour of such evaluations of specific measures, particularly those protecting habitats, do seem to be increasing, fostered by policies such as United Nations General Assembly (UNGA) Resolution 61/105 in 2006, and subsequent adoption of the FAO International Guidelines for the Management of Deep-Sea Fisheries in the High Seas in 2008, which hold fisheries authorities accountable for ensuring fishing activities do not result in SAIs on deep-sea VMEs.

When ABFMs are granted OECM status, each area, with the specific measures applying to it, ought to be assessed 'case by case', individually and recurrently, with undefined frequency but in the long term, to demonstrate the continuous production of expected biodiversity benefits. The need for monitoring of an OECM performance is stressed in Decision 14/8 (Annex III, C, 1, f). The Decision Annexes are not mandatory but it should be stressed that the mandatory definition of OECMs in that Decision states that the OECM is expected *to achieve positive and sustained long-term outcomes* and that providing evidence that this is the case requires some form of monitoring and performance evaluation through a more or less sophisticated process. The frequency of these evaluations is a decision of the State and may depend on the biodiversity of concern, the means available in the fishery MER, etc. However, for OECMs, evaluations of effectiveness are expected to be reported to UNEP-WCMC in the long term, in the global OECM database, and the WCMC Secretariat is committed to ask States for an update of their OECM information every 5 years (UNEP-WCMC, 2019) (see Section 7.5.2).

In modern fisheries (and indeed in traditionally managed ones), the monitoring of fishery resources and associated biodiversity is undertaken together, for logical and operational reasons. However, in the following sections we consider the MER undertaken for the fishery as a whole to assess its performance (referred to as fishery MER) separately from that undertaken specifically for the OECM (referred to as OECM MER). This approach distinguishes the two functions to illustrate the differences, complementarities and synergies possible between them, but in practice, the two will probably be operationally integrated in a single process (see Section 7.1.3).

7.1.1 The fishery MER

In order to track the broad performance of the fisheries and their management and inform the legitimate authority accordingly, the fishery MER activities aim to collect diversified information from the fishery actors and dedicated scientific surveys and analyses. Some or all the tasks may be undertaken by a fisheries research laboratory under the State institution in charge of fisheries and sometimes also of the environment. It usually involves a range of fishery offices (e.g. in charge of fishery statistics, control, surveillance) as well as other institutions such as environmental research institutes and universities and, particularly in community-managed fisheries, community members themselves and NGOs.

At national or regional levels, the way fisheries are monitored and assessed for performance depends on the resources, including personnel, made available to undertake the tasks. The amount of resources and people invested in monitoring, evaluation and reporting of specific fisheries depends *inter alia* on the wealth of countries involved; the number, size and socioeconomic importance of the fisheries concerned; the level of development and breadth of the fishery research support; and the fishery management approach adopted. Except when specifically mentioned, Section 7.1 refers to monitoring systems operating in well-managed fisheries, with reasonably functional MER processes. These processes may be adapted in capacity-limited situations (particularly, but not exclusively, in small-scale, community-managed fisheries), in ways consistent with the spirit and expectations of a fishery MER. These adaptations can help to manage costs but have implications for robustness and credibility of the conclusions, and the simplified processes should ensure transparency with regard to the information used and the rationales for their actions and conclusions.

The fishery MER activities aim to provide timely information to the legitimate authority on (i) fishing activity and outputs; (ii) status and trends of target species; (iii) implementation of management measures and level of compliance; (iv) collateral impact of the fishery on dependent and associated species, particularly protected species, and on vulnerable habitats; and (v) selected aspects of the ocean and climatic environment of importance for the understanding of resources movement and productivity, fishing operations dynamics and management performance. More specifically, the function aims to checks whether, for example: (i) the fishery management plan is being implemented as effectively and efficiently as designed; (ii) the expected outcomes of management measures are being achieved or can reasonably be expected to be achieved; (iii) emerging or unanticipated issues arising that could affect performance, and to what extent; and (iv) what possible options exist to cost-effectively improve performance in (i) and (ii), and to better prepare for (iii).

Diverse potential socioeconomic frameworks may be used to monitor and assess performance of fisheries management measures, e.g. with relation to social and economic costs and benefits.[2] Different evaluation frameworks place different emphasis on the diverse social and economic aspects of a fishery performance. Consequently, the most suitable framework may depend on the priorities among the social and economic objectives, which in turn may differ among the stakeholders, rights-holders and various levels of government.

The indicators and hence the data to collect are also highly diverse and situation specific, covering a range of costs, benefits and incentives of financial or social nature. The ecological economics and bioeconomic frameworks provide bases for including conventionally recognized environmental costs and benefits in an integrated analysis of performance and the trade-offs involved in various conservation approaches. The success of such frameworks hinges on the ability to attach 'values' to the range of environmental costs and benefits, properties that can be contentious in their own right (IPBES, 2022b). The role of social sciences is central in assessment of issues relating to equity, livelihoods, social and cultural values, human rights and environmental stewardship. A high level of participation is necessary at different scales: individual, household, community, company, fishery and State levels.

The individual activities of a fishery MER can be described as follows.

- *Monitoring*: to regularly collect data and information to allow the assessment (see next point) of the state of living resources used and impacted by fisheries, the natural and anthropogenic pressures that affect them, and the effectiveness of the fishery management process (e.g. measures, costs, compliance). This requires measuring change in key indicators of: (i) fishing activity and pressure (e.g. areas fished, timing, gear used, fishing effort); (ii) fishery outputs (e.g. amounts and composition of catch, landings, by-catch and discards); (iii) environmental conditions that might also affect the resources (e.g. temperature, rainfall, river outflows, climate change, environmental degradation); and (iv) management measures. In general, the cost-effectiveness of the management system, an important factor in the long term, is

not routinely monitored. Monitoring may be achieved using fishery-dependent data (e.g. on catch amounts and composition) as well as fishery-independent data (e.g. collected through scientific surveys). Effective monitoring of these elements requires early identification of informative, affordable and robust 'indicators', as well as an effective system of data archiving and management, and methods for ensuring monitoring results can be communicated to a wide range of stakeholders (see Section 7.6). Moreover, data are also collected about the MER activity, data collected, costs, outcomes, etc. to audit the fishery MER performance (see Section 7.7).

- *Evaluation*: to systematically analyse monitoring records, to: (i) develop an understanding of the status and trends of the living resources used and impacted by fisheries; (ii) provide a measure of progress towards fisheries management and conservation objectives and targets, with related indicators and performance measures; and (iii) improve management foresight on emerging threats. A key challenge is identifying and accounting for uncertainties in the assessments and how those uncertainties contribute to risks in the management advice. Usually on longer cycles, but ideally still at regular intervals, evaluation of the performance of the overall FMP, of individual spatial and non-spatial measures in the FMP, and appropriateness of the fisheries and conservation objectives and targets should be conducted. In addition, the performance of the fishery MER process itself, in terms of meeting the objectives assigned to it, may be evaluated (see Section 7.3.2).

- *Reporting*: to inform: (i) the legitimate management authorities for development and management decision making, raising of internal awareness of the management challenges, successes and uncertainties and proposing eventual corrective measures for adaptive management; (ii) the fishery sector and other stakeholders, including the co-operating agencies and funding bodies; and (iii) the public at large, as part of the broader public accountability regarding the state of the resources and the management actions, costs and benefits.

The reports may also go to the internal or external audit offices, for use in future evaluations of performance (see Section 7.7). Key feedback expected from the legitimate authority, after it has considered the report, its conclusions and advice, could be in terms of corrected or new objectives (co-ordinating fisheries management and biodiversity conservation), new targets as appropriate, and new measures or adaptation of existing ones.

These three main tasks require a continuous or occasional capacity to collect statistics, deploy observers at sea, use remote sensing technology, and use assessment methods and models to assess the ecological, economic and social conditions of the fishery and its resources in the competence area. Participation of key stakeholders and knowledge holders in the establishment and functioning of the MER activities is an asset, and has additional value in capacity-limited situations such as small-scale fisheries and remote coastal communities. Ideally, the description of the fishery MER process and methods should be well documented, as part of the documentation on the fishery management system and specific fishery management plans (see Section 7.6).

An abundant literature is already available on monitoring of fisheries and conservation, for example for MPAs in Hockings (1998), Pomeroy et al. (2004, 2005), Fancy et al. (2008), Field et al. (2004, 2005, 2007), FAO (2003, 2009a, b), Cochrane and Garcia (2009), and Lindenmayer and Likens (2010)[3] in which more operational guidance can be found.

7.1.2 The OECM MER

To maintain its international status and hence be accounted for in international conservation coverage targets, the OECM is expected to continue to meet the definition and identification criteria adopted in Decision 14/8 regarding the biodiversity benefits and additional properties (see Section 5.5). This is most readily demonstrated when these required properties are translated into management objectives and targets, specific to the OECM but integrated in the fishery management plan (see Section 6.1). Indicators of pressure, status and trends of relevance to

the biodiversity attributes of concern are then reviewed and evaluated to provide evidence of OECM performance (see Section 7.3), inform its adaptive management and report to the CBD and WCMC.

A good part of the background information needed to develop the OECM MER programme[4] is likely to have been already compiled during the identification process (see Chapter 5) and may indeed be provided by the pre-existing fishery MER. However, additional information is likely to be needed to better address the specific biodiversity attributes of concern and other locally relevant values of the OECM. This may require: (i) additional monitoring or analyses added to the workplans for the standard fishery MER activities (e.g. on-board sampling programme; scientific surveys cruises, special working groups; collaborations; governance process; reports; (ii) additional information collected directly by the OECM MER using resources allocated to the OECM MER for such tasks or obtained through collaborations with partner agencies. The new information generated by the OECM MER may be stored and managed by the OECM MER for its own purposes, but should also be included in the data and information system of the fishery MER for a wide range of possible uses. Conversely, the desire to address future threats to biodiversity (including climate change) and issues of representativeness and connectivity (networking) of OECMs may lead to a need for additional monitoring data and tasks at higher geographical scale, in the fishery MER itself and at EEZ or ecosystem level (see Chapter 6 on integration).

As mentioned in the introduction to this chapter, the OECM MER and fishery MER are distinguished here, despite their interconnections, because the OECM status of an ABFM makes it necessary to report separately on its performance in relation to the biodiversity features of concern. In practice, however, the two MER activities need to be integrated because of numerous financial, operational and ecological reasons, including the following.

- Both MER processes use similar services: the MCS system, to ensure compliance; the statistical office, to collect, process and maintain fishery statistics; the research vessel, to collect fishery-independent data; and the fishery research laboratory, university and environmental agencies, to undertake multidisciplinary assessments. This makes common sampling and analysis the most efficient overall approach.

- Many biodiversity attributes of concern in the OECM are likely to be also present, move or migrate outside it, and be interconnected through the trophic web. Consequently, the biodiversity outcomes of the OECM may spill over to the surrounding ecosystem and be affected by what happens outside the OECM, calling for integrated assessments;
- Collection of comparable data inside and outside the OECM would inform assessment of the level of protection provided by the OECM as well as the level of impact of the fishery.
- Even though the biodiversity conservation objectives are fundamental for OECMs, the ABFM to which the OECM status has been granted keeps its fishery sustainability objectives which consequently remain a major concern for the fishery MER.
- In an EAF approach, fisheries management already applies conservation measures for non-target species and protected species (e.g. gear regulations, by-catch excluder devices, by-catch quotas, seasonal or permanent closures), calling for integrated assessment of conservation performance.
- The two MER processes report to the same legitimate authority, preferably through the same channels, providing an additional and strong incentive for their effective integration.
- The fishery MER and OECM MER processes are confronted with many common external drivers and pressures (including climate change) and seeking similar outcomes for status, trends and performance.
- Because of these numerous interactions, the fishery MER may need to be strengthened to respond to the new or reinforced biodiversity conservation objectives of the OECMs.

As a consequence, it would be logical to integrate the fishery MER and OECM MER activities, just as OECMs are logically integrated in the FMP (see Section 6.1), even though the respective functions are different (albeit connected and complementary) with different legal bases, obligations, reporting periodicity and institutional recipients. The higher the complementarity and spatial overlapping of the fishery MER and OECM MER (e.g. in terms of objectives, targets, management activity and resources), the more comparable the two MERs are likely to be. For efficiency and consistent with the EAF approach, the OECM MER may be conceived as a component of the fishery MER with specific reporting requirements. For the same reasons, fishery MER systems covering various fisheries operating in the same 'ecosystem', EEZ or transboundary area, or MERs developed in other economic sectors that have spatial or functional areas of overlap, would similarly benefit from co-ordinating information exchange, and some level of integration in its analysis and interpretation.

In the OECM MER, more focused attention is given to those tasks and elements of biodiversity that are additional to those that would have been expected by the fishery MER from the original ABFM in terms of sustainability of target species use and eventual broader conservation concerns. The new needs depend on how well the existing fishery MER activities, possibly under an EAF approach, were already covering broader elements of biodiversity such as non-target species, including threatened and protected species, essential habitats and other biodiversity attributes of concern. The OECM performance in terms of equitable governance might be facilitated in modern fisheries where principles of good governance are already applied unless local adjustments are needed, e.g. in fisheries across overlapping jurisdictions.

The actions required for developing and running an OECM MER within the fishery MER are considered in some detail in the following section in relation to three domains: (i) strategic planning and co-ordination; (ii) monitoring and evaluation; and (iii) data and information management.

7.2 Strategic Planning of the OECM MER

Strategic planning turns a vision (of a desired future) into a strategy with broad goals, specific objectives, activities, resources for implementation and, possibly, the mechanisms to guide and control the implementation, such as governance, oversight, management, reporting and audit. In the case of an OECM MER, many of these elements could be 'inherited' from the pre-existing ABFM governance, planning and implementation, but may need to be tailored for the requirements of OECM status.

The OECM definition provides the overarching goal of fishery OECMs:

> to achieve positive and sustained long-term outcomes for the *in situ* conservation of biodiversity, with associated ecosystem functions and services and, where applicable, cultural, spiritual, socioeconomic and other locally relevant values.

As a consequence, more specific biodiversity conservation objectives may need to be set for the specific OECM site, regarding factors such as: (i) performance of the site and its specific measures in terms of status and trends of the biodiversity of concern and other values; and (ii) performance of the OECM MER process itself in providing the necessary information. Both objectives may be operationalized by targets, indicators and reference values (or benchmarks) to facilitate the performance assessment.

The specific upstream activities needed to plan the work of an OECM MER include the following.

7.2.1 Describing the types of outputs expected from the OECM MER on biodiversity conservation

This will help efficient planning of the MER biodiversity conservation-related activities, in addition to the activities regarding the fishery sustainability itself.[5] The main outputs relate to the assessment of the OECM performance in meeting the biodiversity conservation objectives that justify the OECM status.[6] A second important output is an assessment of the performance of the OECM MER in delivering its MER tasks. Both require monitoring and evaluation of partly overlapping elements of the fishery and the ecosystem in which it operates. The MER report to the legitimate authority on OECM performance is expected to contain:

(i) data and information on the evolution of fishing operations and other drivers, including external drivers if available; (ii) the evolution of status and trends of the biodiversity attributes of concern, including the relevant ecosystem services; on the benefits (including harm reduction), costs and their distribution among stakeholders (noting that evaluating 'benefits' may require comparative information on the status and trend of the biodiversity attributes of concern outside the OECM); (iii) the evolution of external drivers (e.g. the global economy, markets, climate change, price of fuel) and early warnings on impending threats, if any; and (iv) based on the above, synthetic conclusions on the performance (effectiveness and efficiency) of the measures taken, in the OECM and its surroundings, with eventual considerations or recommendations on mitigation measures.

7.2.2 Describing the specific types of tasks assigned to the OECM MER

The expected outputs determine the types of tasks, e.g. (i) collecting information on the biodiversity attributes of concern and other relevant values as well as fishing activities and other pressures and threats, in the OECM and around it (in co-ordination with the fishery MER); (ii) assessing their status and trends in relation to the specific conservation objectives and related indicators and reference values; and (iii) elaborating management options to maintain good trajectories or correct unsatisfactory ones, based on known or assumed drivers; and (iv) assessing the performance of the OECM MER in meeting conventional standards of effectiveness, efficiency, timeliness, relevance, accuracy and treatment of uncertainty.

7.2.3 Documenting the additional biodiversity elements to monitor and evaluate

These elements might include: (i) biodiversity attributes of concern such as vulnerable species, communities and habitats; areas important for life cycles, ecological representativeness and connectivity; ecosystem functions and services including food and livelihoods; (ii) other features of social, economic, cultural and spiritual importance; (iii) current and projected fishing pressure; (iv) other relevant pressures and threats, with their degree of significance and likelihood, requiring information from non-fishery sources; and (v) governance ensuring equity in the identification of relevant participants, their involvement in the governance process, and the distribution of costs and benefits. These elements (including their historical review) should have been identified and documented during the OECM identification process (see Chapter 5). They may only need to be confirmed in the initial OECM MER documentation and, as necessary, updated as the OECM and its drivers evolve. The elements to monitor are numerous and the budgets are limited. Context-sensitive priorities may therefore need to be explicitly established, based on the relative ecological, social (cultural) and economic importance of the elements, and feasibility of monitoring (considering both sensitivity and specificity of indicators), resulting in a subset of elements being highlighted as 'key performance elements' for the OECM.[7]

7.2.4   Identifying the additional objectives and targets of biodiversity conservation

For each of the biodiversity attributes of concern identified above, the specific objectives and related targets need to be credible for both the fishery and conservation communities. They need to be enacted at management level, e.g. when integrating OECMs into the management plan (see Section 6.1). They also need to be specified in the fishery MER programme. If additional ones are identified in the OECM MER process, they will need to be endorsed formally by the legitimate authority and retrofitted in the fishery MER programme and the FMP.

There are several possible sources for setting conservation benchmarks for biodiversity features. Following the Biodiversity Impact Mitigation (BIM) hierarchy[8] (Wilcox and Donlan, 2007; ten Kate and Crowe, 2014; Milner-Gulland et al., 2018), the management objective may be to avoid, minimize, mitigate/remediate or compensate (where possible) a

specific impact on biodiversity, to either maintain status quo (No Net Loss objective) or restore the element to some reference state (Net Gain objective). Following the Law of the Sea, the reference state for dependent and associated species might be the biomass *below which their reproduction would be threatened*[9] (UNCLOS Art. 61.4), whereas Annex III of UNCED specifies the generic precautionary standard for all environmental features as 'avoidance of serious or irreversible harm'. With very similar implications on biomass, the CBD, in Target 6, requires that these species *should not be affected by Significant Adverse Impact (SAI)*, a benchmark established in UNGA Resolution 61/105/.

7.2.5   Identifying indicator baselines, reference values and priorities

Indicators and their benchmarks may be qualitative or quantitative and correspond to each objective and biodiversity attributes of concern. Most of these might have been determined during the identification process (see Chapter 5). Factors to consider include: (i) priority given by the legitimate authority to the various objectives; (ii) cost of data acquisition compared to budgetary resources available; (iii) data versatility (usability across many objectives and OECMs); (iv) complexity of the indicator elaboration pathway; (v) ability to communicate the features' status and change; (vi) sensitivity and specificity of indicators compared to that needed for effective dynamic management (signal/noise ratio); and (vii) support and trust of stakeholders. A large literature is available on the use of indicators for policy development and management in both fisheries and conservation (see Pomeroy et al., 2005; Rice and Rochet, 2005; Garcia et al., 2009; Addison et al., 2018). Whatever the indicators chosen, the performance measure can be the ratio between the observed value of the indicator in the fishery-OECM compared to the reference value for that indictor in well-functioning populations or communities of the same type, in the same areas sometimes in the past or, when known, the values that correspond to healthy or precautionary benchmarks as defined in the agreements cited in the previous section on identifying the additional biodiversity conservation objectives and targets. The indicators could conveniently be organized along the pressure-state-response (PSR) framework and its variants (Moldan *et al.*, 1997; Chesson, 2013) and the MER would track the current pressures and impending threats (P) on the status and trends (S) of assets, and the management responses (R), clarifying cause–effect relationships between P, S and R.

7.2.6 Listing additional management measures applying in the OECM

Some of the measures to be applied in the OECM might have already been in place in the ABFM before becoming an OECM and should be integrated into the overall FMP and monitored. Additional measures, or modification of existing measures, may be needed to reach the additional biodiversity conservation objectives. These measures might have already been considered during the identification process and proposed to the unit of the legitimate authority in charge of management.

A wide range of modifications or additions to the fishery measures could be considered, such as, *inter alia*: (i) banning a damaging gear or practice in the OECM; (ii) changing the gear specifications to modify selectivity and avoid by-catch of particular species or reduce by-catch overall; (iii) changing the seasonality of the fishery in the OECM, to avoid the timing of particular life history activities (breeding or calving area, etc.), and many other measures. The measure of performance (in terms of harm reduction) for each type might, respectively, be: (i) confirmation that a forbidden gear or practice has been eliminated (and no alternative source of mortality was introduced); (ii) the reduction of by-catch of the species of concern; or (iii) reduction in perturbation of the life history feature of concern. It should be noted that when species of concern may also be taken by the same or another fishery outside the OECM, as the species move or migrate in and out of the OECM, co-ordinated by-catch reduction measures might be needed outside the OECM, if by-catches occurred on a scale that would pose a risk of reducing or losing the net conservation benefits.[10] Such outside measures are not part

of the OECM management 'responsibility' but should be considered by the fishery MER and the management authority to improve the overall biodiversity conservation performance in the fishery.

7.2.7 Listing the elements to monitor in order to assess OECM MER performance

It is important to check not only whether the OECM MER is effective (obtaining the expected outcomes) but also efficient (in obtaining them cost-effectively). Consequently, time series of records should also be collected concerning the OECM-related investments in management, e.g. (i) collection of the broad range of monitoring records, processing and managing them; (ii) assessment of status and trends (research costs); (iii) control and surveillance, to assess compliance with the regulations applying into the OECM; and (iv) administrative and other costs of running the OECM MER. These costs might be directly supported by the OECM MER budget – probably a subset of the fishery MER budget – but may also be incurred by collaborating services and agencies contributing to the OECM MER. In well-managed fisheries the incremental cost of the OECM MER should be only a modest increase of the overall fishery MER activities unless the need to fully document the occurrence and trends in specific biodiversity benefits requires additional indicators that are expensive to collect (e.g. having to conduct frequent aerial surveys for rare but highly vulnerable cetaceans).

7.2.8 Selection of monitoring and assessment methods specific to the OECM MER (if any) to describe the change in status of biodiversity attributes of concern or pressures exerted on them

Again, many of the methods used in the process of identification of OECMs to assess pressures, threats, risks and biodiversity benefits are likely to be used also for the recurrent assessments conducted in the fishery-OECM (see Section 7.3). The methodology could improve with time as the MER activities collect more data, master

new methodologies or acquire new competences. The flexibility foreseen in Decision 14/8
regarding the identification process also applies
mutatis mutandis in the MER functioning, for the
same reasons, including objective limitations in
the data and capacity available. The methods to
be used in each national and ecological context
need to be explicitly defined, e.g. as Standard
Operating Procedures (SOPs), in order to
ensure a level of standardization, consistency
and coherence across fisheries, time and space,
in data collection, processing, analysis and
interpretation of changes in the indicators. In
areas where capacity is sufficient, the management strategy evaluation (MSE) process used in
advanced fisheries management programmes,
modified to deal with broader biodiversity values
(see Smith *et al.*, 2007), could be used to test
the robustness of SOPs to uncertainties in data,
assessment methods and decision processes.
Methods for quick or in-depth assessments,
using local knowledge, expert views or sophisticated simulations, may be combined to deal
with different attributes or varying budgets (to
optimize costs) or account for sensitivities or
needs of the evaluators and decision makers.

7.2.9   Identifying and strengthening competence collaborations

Possible activities in that direction include: (i)
identifying the potential participants of a collaborative process; (ii) establishing formal collaborations with institutions and organizations
monitoring the ocean environment, biodiversity
and social and economic parameters; (iii) identifying additional sources of data and assessment
competence (e.g. on seabirds, marine mammals,
turtles, seahorses, snakes, depending on objectives and expected biodiversity outcomes) and as
partners in MPA and OECM networks (increasing mutual trust). These collaborations should
be taken into account in the fishery MER.

7.2.10   Describing the types of outputs expected in the MER report on the biodiversity attributes of concern[11]

Identifying the specific outputs expected from
the OECM MER, in its three major programme
areas (monitoring, evaluation and reporting),
on the situation and performance of the OECM
and the OECM MER, will help efficient planning
of MER activities in these areas. The MER report
to the legitimate authority is expected to contain
all the information described earlier in Section
7.2. Any conclusions and recommendations
should be drafted to facilitate the preparation
of the report to be submitted to the WCMC. A
separate report might also provide an appraisal
of the performance of the MER programme
itself (self-evaluation) with suggestions for
improvements.

Having undertaken the upstream activities indicated above, the MER activities may
be undertaken as suggested below (sections
7.3–7.5).

7.3   Monitoring and Evaluation of Performance

As stressed earlier, two aspects of performance
need to be considered: (i) the performance of the
OECM area (with the measures applied into it) in
delivering the expected biodiversity outcomes;
and (ii) the performance of the OECM MER
system or programme in discharging its monitoring, evaluation and reporting tasks. These
two interconnected aspects are examined below.

Because 'effectiveness' is an important
point in Decision 14/8, the consent of the legitimate authorities (including FPIC) that is needed
for the identification process should include a
commitment to monitor and evaluate the OECM
performance in the longer term, with the means
available and, as appropriate, support from the
State and/or collaborating institutions.

7.3.1 Ongoing assessment of the OECM's effectiveness

A fishery-OECM is expected to produce its biodiversity outcomes in the long term and its performance in doing so should therefore be recurrently assessed, in terms of effectiveness (i.e. its capacity to inform management on the extent to which it reaches its fishery and conservation objectives and targets) and efficiency (i.e. its capacity to do so cost-effectively). The issue of effectiveness is covered in Section 7.4. The issue of efficiency is not explicitly addressed in Decision 14/8 but the issues of 'costs', 'benefits' and 'costs and benefits sharing' are abundantly referred to in the Decision, in relation to equity (see Section 7.4.2).

An OECM's effectiveness may be recurrently evaluated by the extent to which (i) its biodiversity outcomes match its stated objectives, and (ii) the different actors are satisfied with them. Only the outcomes related to biodiversity, habitats and ecosystem functions and services directly determine maintenance of the OECM status. The other types of outcomes may strengthen or weaken the evaluation and indicate areas for improvement of the OECM without affecting its status.

The effectiveness evaluation process is more complex than just reviewing trends in the biodiversity features of concern identified in the OECM objectives, because it should be shown that the outcomes are probably caused by the OECM measures. In reality, many factors other than fishery operations can influence the outcomes, including external environmental and socioeconomic drivers, mediating governance and management factors and measures, and the characteristics of use (see Section 7.4.4). Moreover, many fishery-OECMs will have objectives related to several biodiversity features of concern, and even under favourable conditions, it is unlikely that equal progress can be made at the same rate on all of them. Finally, effectiveness is also unavoidably perceived differently by different groups of actors with different interests.

Performance of the OECM necessarily reflects the quality of: (i) the OECM area location and size, and the fit between the area boundary and the distribution and movements of the biodiversity attributes of concern it is intended to protect; (ii) the technical measures applied in the OECM area and, possibly, around it (e.g. access rules, gear regulations, fishing practices); and (iii) the degree of participation of the actors directly impacted by the measure and the related enforcement effectiveness. The data and analyses required to assess performance are similar to those undertaken during the initial identification process (see Section 5.5), albeit more focused and with more recent and specific data, and the conclusions of the identification could be updated.

Approaches to performance assessment have been developed and tested in fishery and conservation science for many types of populations and ecosystem features (see Section 5.2.8) and can be used as appropriate for the various contexts. The assessments must be conducted in a timely manner, facilitating a rapid management response or fulfilling the commitment to legislated reporting frameworks. The capacity to respond also to *ad hoc* questions from the authority (e.g. in case of unexpected events) and others engaged in the governance processes would also be essential.

The tasks necessary for a robust performance evaluation rely on the information collected during monitoring and include the following.

7.3.1.1 Selecting indicators

For each OECM objective and target, a range of indicators might be considered. A selection could be made based on their: (i) measurability; (ii) affordability, i.e. data acquisition costs, relative to budgets available; (iii) versatility, i.e. usability across many objectives or OECMs; (iv) elaboration complexity, relative to research capacity available; (v) sensitivity, accuracy and precision, relative to that needed for effective dynamic management (including signal/noise ratio); (vi) specificity of the relation between the changes in the indicators, in the attribute of concern, and the causes of the changes; and (vii) ease of communication, and trust/support of stakeholders. A large literature exists on the use of indicators of fisheries sustainability (e.g. Chesson and Clayton, 1998; FAO, 1999; Rice and Rochet, 2005; Garcia *et al.*, 2009; Anderson *et al.*, 2015) and for biodiversity conservation (BIONET and IUCN, 1997; Ablan *et al.*, 2004; CBD, 2004; Pomeroy *et al*, 2004, 2005; Biodiversity Indicators Partnership, 2010, 2011). These sources can be used to select the best practices corresponding to the data and competences available, and the needs of those responding to the indictors (legitimate authority, other participants in governance, other collaborators). Choice of methods should also focus on the biodiversity properties claimed to be addressed in the OECM, and the types of outcomes it is expected to produce.

7.3.1.2   Analysing trends

Analysing trends in: (i) fishing operations as the main source of pressure affecting the various elements of concern, disaggregated to inside and around the OECM; (ii) external drivers such as global economy parameters of relevance, market demands, climate change, fuel prices and their observed or potential impact on the elements being assessed; (iii) impending threats from the fishery sector and as much as possible from other sectors[12] and related potential impacts; and (iv) status of biodiversity and other values of concern, including ecosystem services, updating the archived historical and ongoing records and improving the ability to take into account causal relationships between changes in pressures and biodiversity values.

Accurately capturing trends and their changes is important to assess the effectiveness of a measure and adjust it if necessary. To be able to capture trends and their changes, it is essential to have a measure of the 'natural' variability of indicators, to allow reliable detection of signals of change (e.g. due to the introduction or change of a measure) from the ambient 'noise'. This is not easy to achieve in complex ecosystems with numerous interacting variables. The challenge represented by increasing indicator variance in increasingly complex systems may be partially mitigated by the fairly standard practice of linking the benchmarks for an indicator to the cumulative frequency of past observations (or modelled values) of the indicator, rather than using fixed percentages of the indicator mean, regardless of indicator variability.

However, showing evidence of a positive biodiversity outcome then becomes very difficult for benefits measured by indicators with substantial variance. In such cases, the benchmark for an *in situ* minimum detectable benefit would have to be a very significant improvement of the biodiversity feature, e.g. statistically making it two standard deviations or more above the mean of the indicator prior to the OECM management measure(s) being implemented.

7.3.1.3   Assessing OECM performance with regard to biodiversity conservation

This is a central product of the fishery-OECM MER, needed to verify that the OECM status of the area has been maintained. Performance is assessed by comparing the status and trends of the biodiversity attributes of concern to the benchmarks for related objectives and targets. If the match is good, the measures in the OECM might be operating as expected. If the status and trends fall short of expectations, the measure may need to be changed or augmented. If the trend is right but slower than expected, the measure may need to be strengthened or the reaction time might have been underestimated. Whether such delays affect OECM status may depend on how degraded the biodiversity feature was before the OECM status was attained, and how effectively the measures can strengthened.

Indicator-based metrics of apparent success and failure of a management or conservation measure may be due entirely or in part to a variable environment. In addition, the fishery-OECM performance is related to multiple indicators corresponding to the different biodiversity values of concern, and their trends are unlikely to be perfectly synchronized, for example because of life history differences among biodiversity taxa. Consequently, definitive conclusions on performance will require sufficient time for the cause–effect relations to play out, making it important to take life history considerations into account when setting goals for population recovery or minimum status. Comparison with similar types of OECMs elsewhere is the only way to evaluate management performance in systems with moderate to high natural variation. The conclusions reached in the OECM, eventually complemented by conclusions obtained in the fishery MER, should show how effective management is in maintaining or improving the biodiversity attributes and other values of concern and in controlling fishing pressure in the OECM area and around it.

7.3.1.4 Assessing the 'additional' properties of the OECM

These properties were described in Section 5.5.5 and refer to: (i) ecological representativeness; (ii) connectivity; (iii) complementarity; and (iv) integration. They may enhance the OECM effectiveness and strengthen, but are not essential in, the justification for an OECM status. Hence, they can affect the improvement of the OECM performance but do not threaten their status unless these additional properties were included in the OECM rationale, and they are found to be degrading within the OECM. Properties (i) to (iii) relate to relations between the biodiversity inside and outside the OECM, and may change with time as biodiversity evolves and knowledge improves. Property (iv) may change if the integration of the OECM in the management plan and the OECM MER changes. The effectiveness of this integration would also be an argument for assessment of the OECM MER performance (see Section 7.3.2).

7.3.1.5 Elaborating options for new or improved measures

Elaborating options for new or improved measures in the fishery-OECM, or around it, is essential in case of unsatisfactory performance, in order to improve it. If unsatisfactory performance is due to failure to reduce pressures or enhance productivity, additional measures for effective harm reduction and/or faster recovery would be necessary to maintain OECM status. If the unsatisfactory performance is because of excessive costs or inequities in distribution of costs and benefits among subsectors and coastal community groups, the legitimate authority may demand improvements in governance and/or use of less resource-demanding measures to manage the pressures, as appropriate, but OECM status is less likely to be in imminent jeopardy. Depending on the governance system, the OECM MER may be asked to recommend a particular improvement, with the rationale for its selection.

The activities above are presented in sequence but they are likely to be undertaken in parallel, interacting with each other. The results might have highly technical components which are needed to design effective adaptive responses to shortcomings. However, communication with the many interested perspectives is aided with simpler tools to also express the results in readily comprehensible ways. For a more intuitive understanding, the results can be used to elaborate a dashboard for the OECM, such as integrated in the fishery MER dashboard (see Fig. 7.1 and its caption for explanations, and Garcia *et al.* (2009) for more details and illustrations).

7.3.2 Assessment of the OECM MER

The performance of the OECM MER programme or activities can be judged by the extent to which it functions according to operational objectives and plans when: (i) collecting the data needed for the assessment of the OECM and OECM MER performance; (ii) undertaking the assessments required for both, including responsiveness to unexpected situations; and (iii) timely and accurate reporting on both to the legitimate authority (see Section 5.6.2) with adequate recommendations for its improvements through adaptive management. The performance of the OECM MER reflects necessarily on the performance of the fishery MER into which it is integrated and, by extension, on the performance of the whole FMP and, increasingly often, the performance of the more integrated cross-sectoral ecosystem management goals. The assessment would be part and parcel of the activity report of the OECM MER, probably integrated in the activity report of the fishery MER to which it is connected. Such a report would contain a detailed compilation of, for example, the scientific surveys, sampling plan and other data collection activities; data processing and storage; assessment activities[13] including working groups; methodological developments in data collection and assessments; collaborations and participation; funding and expenses. The comparison between planned and effectively undertaken activities will provide an assessment of performance of the monitoring system.

Many elements condition the performance of the OECM MER, as for that of any MER, including: (i) the quality of its sampling

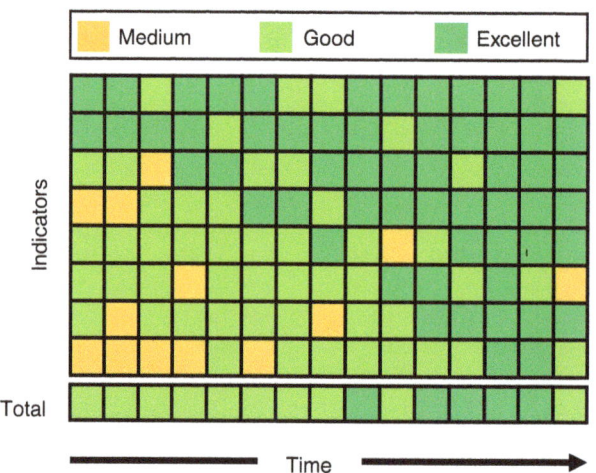

Fig. 7.1. Illustrative evolution of a set of indicators of biodiversity values and/or pressures in an OECM, using a standard colour coding for medium (yellow), good (light green) and excellent (dark green) performance. In this example, the overall trend is positive as the proportion of green 'scores' increases with time. An opposite trend would call for corrective measures.

programme, its continuous optimization and adaptation over time; (ii) access to modern technology (e.g. research vessels, remote sensing, digital mapping, underwater video, data management, assessment software); (iii) its success in establishing institutional collaborations and active participation of stakeholders, e.g. in collection of information and assessments and communication; (iv) the quality of its reports, their accuracy, timeliness and accessibility to all stakeholders; (v) the quality (safety, durability) of its data and information management system; and (vi) its responsiveness to unexpected events (particularly negative events) and auditing recommendations about shortcomings in the OECM MER. These issues are not different in kind from those for a fishery MER itself, and often must be resolved at that level. Where the sources of the shortcomings are rooted in governance challenges or the activities of other sectors, more partners from the relevant agencies or groups are needed to find pathways to solutions.

The performance of the OECM MER and that of the fishery MER are interdependent to some extent as the two programmes necessarily share financial, technical and human resources, produce overlapping or complementary outputs, and their respective contribution to ecosystem-level outcomes can be hard to

separate.[14] Moreover, OECM MER activities are organized differently in different countries. Some keep the large majority of fishery MER and OECM MER expertise and resources within the same agency, others have portions of the necessary resources and expertise (particularly for biodiversity) in other departments of the State or different levels of government, whereas others obtain some or part of the expertise from academics, ENGOs or private sector companies. It is therefore important to clearly define the specific responsibilities of the OECM MER activities in order to correctly measure its overall performance, apply the needed corrective measures and ensure there is appropriate evaluation of the performance of all other partners in the management of the OECM, and delivery and evaluation of its outcomes.

The performance of the OECM MER can be assessed internally (self-evaluation, internal audit) and could be supplemented by an occasional independent third-party auditing process, if so desired. The OECM MER may be assessed alone but for practical reasons should usually be assessed together with the fishery MER with which it is integrated, unless partners in delivery of the OECM outcomes are not usually engaged in the fishery MER.

7.4   Evaluation of Fishery-OECM Effectiveness[15]

A central function of an MER is to evaluate the effectiveness (also referred to as 'performance') of the fishery and hence of the management measures. In this section, after some general considerations on effectiveness, we look briefly at the methodology needed, before considering the factors affecting effectiveness and the potential effectiveness of fishery-OECMs.

7.4.1   General considerations on effectiveness

Assessing effectiveness of fishery-OECMs typically will be complex. Diverse sets of information and methods are required to assess its various dimensions, to take account of both the various actors' perspectives and the effects of external drivers. The information available, itself, may be evaluated *ex ante* (e.g. by simulation), *ex post* (with empirical data) or combining both. Assessing effectiveness requires comparing outcomes with objectives (see below). Ocean and resources dynamics and variability are sources of uncertainty that can accelerate, delay or even derail progress towards such objectives, particularly for outcomes expected to materialize outside the OECM (as spillover) and/or quite some time after measures are put in place.

7.4.1.1   Definition of effectiveness

In this book, we use 'effectiveness' and 'performance' as synonyms. As for any management measure, the effectiveness of a fishery-OECM is (i) the extent to which the biodiversity outcomes that the measure was supposed to deliver actually materialized; and (ii) the extent to which the different actors were satisfied by these outcomes.

Regarding the first element, if the OECM goals and objectives had been precisely stated and the expected outcomes were clear, concrete and measurable, an objective evaluation is possible. The second element is more subjective and less straightforward, *inter alia* because the degree of satisfaction with the various outcomes is likely to be different for different actors. The views of the fishery administration and of the conservation authority are crucial with regard to both elements of effectiveness. However, strong dissatisfaction from either the industry participants (e.g. if the costs to deliver the outcomes are greater than they had been led to expect) or the conservation biology community (if the benefits being received were less than expected) would each bring 'effectiveness' into question. However, if all key actors were involved in defining the OECM objectives and expected outcomes, possible controversies may be limited.

7.4.1.2   Requirement for effectiveness

OECMs can maintain their status as long as they remain effective, meeting the OECM criteria, providing the expected biodiversity benefits and meeting the governance and other contextual standards. This requirement for 'effectiveness' is implicit in the Decision 14/8 definition of OECMs, in the requirement of *long-term (positive) outcomes*, and also explicit in criterion C (see Section 5.5.3). The resulting technical necessity is that the OECM performance in meeting its conservation objectives and targets (possibly in a cost-effective manner) needs to be recurrently checked through the fishery monitoring and evaluation system. In the process, many questions emerge such as: What makes an OECM 'effective'? Who decides on the objectives against which effectiveness is measured? Who judges it? What are the effectiveness criteria? What factors affect effectiveness? At what scale is it appraised? What methods can be used? How much effectiveness is enough to deserve the fishery-OECM status? How equitable is the distribution of management costs and benefits? And what do we currently know about the effectiveness of fishery-OECMs?

Effectiveness is necessarily related to the state and trend of biodiversity attributes of concern but also to the human communities in and around them, to ensure buy-in and compliance. It may be noted that the effectiveness of precautionary spatial measures that might be necessary to address potential threats can only be empirically demonstrated when such threats materialize.

These aspects are briefly addressed below.

7.4.1.3   The role of objectives

As mentioned earlier, fishery-OECMs usually have two broad objectives: (i) to ensure the conventional sustainability of the fishery and its resources and essential habitats; and (ii) to generate long-term benefits, particularly for the biodiversity features of concern. The first objectives are usually the reason for the existence of the ABFM and effectiveness in achieving them should already be established when considering whether to grant it OECM status.[16] However, failure to achieve the expected sustainability outcomes may lead to cancellation or modification of the ABFM area, location and measures, and the nature of those changes could potentially threaten the appropriateness of the OECM status. Similarly, poor enforcement of the fishery management measures outside the OECM may reduce or impair the OECM effectiveness, another reason for good integration of management inside and outside the OECM (see Section 6.2).

Before its identification and possible enhancement as a fishery-OECM, the ABFM broader biodiversity benefits may have been unintended co-benefits. Decision 14/8 goes further, recommending that following formal identification, co-benefits be identified as intended benefits and formally reflected in management plans as objectives, to strengthen the related monitoring and guide performance assessment. In community-based OECMs, objectives may be framed in ways different from those of fisheries scientists and managers and conventional monitoring capacity may be limited. Therefore, effectiveness should be assessed using approaches where local and community knowledge have a meaningful role.

7.4.1.4   Effectiveness is complex and relative

In a social-ecological system, effectiveness is a complex multidimensional challenge. The outcomes observed may be positive or negative. Benefits and impacts may be intended (e.g. the impact of fishing on target species) or not (e.g. through by-catch). They could also be detected or not. The relationships between actions, drivers and outcomes may be neither linear nor monotonic. Effectiveness is also dynamic, evolving with time and in space because of natural and social variability and cascading effects. For example, as discussed in Section 5.2.2, measures intended to generate positive outcomes for some biodiversity features of concern may result in negative outcomes for other biodiversity features and, if the latter are also features of concern, difficult trade-offs may arise, e.g. through predator–prey relationships. Moreover, any judgement of effectiveness of any fishery-OECM is also relative, depending on how ambitious the objectives and expectations were, and subjective, depending on expectations and values of each perspective making a judgement.

7.4.1.5   Scope of the effectiveness assessment

The long-term positive benefits expected from a fishery-OECM are those produced and measured inside the OECM area. However, the OECM interacts with its surrounding ecosystem through export of propagules (e.g. of eggs, larvae, juveniles), active movements in and out of the OECM by resident species (e.g. looking for food) or highly mobile species migrating seasonally to complete their life cycle. Finally, OECM effectiveness may be affected by external factors.

Consequently, the consequences of the measures applied inside the fishery-OECM area might be evaluated within its boundary, especially for resident species and habitat features, but also outside it for many mobile populations or life stages that may leave the area: (i) in case of spillover,[17] when some of the biodiversity features enhanced in the fishery-OECM area move outside it, grow and reproduce, adding biodiversity benefits; (ii) or when these features become vulnerable around the fishery-OECM, to the fishery in which the fishery-OECM operates, as well as other spatially overlapping fisheries, affecting the magnitude and security of the benefits obtained overall from the OECM. Addressing the problem requires integrating technical measures taken inside the OECM area and outside it (see Chapter 6).

The overall effectiveness of the OECM may also depend to some extent on the impact of other sectors in the OECM (if relevant), in the general area, in the EEZ and in the regional ecosystem or seascape, along with coastal degradation, land-based pollution and climate change.

A key implication is that effectiveness would benefit from good co-ordination to improve coherence, complementarity and synergy between the management of the OECM and that of the whole fishery on biodiversity objectives. For example, if turtles are totally protected in the fishery-OECM, there could also be the need for other spatial measures (like a protected corridor) and non-spatial measures (like mandatory by-catch excluder devices and caps on accidental turtle catch outside the OECM) to avoid dissipating the benefits produced by the OECM.

Moreover, effectiveness of a fishery-OECM network might also be appraised at the level of the EEZ, ecosystem or seascape, together with the MPAs network, assessing their complementarity and joint contribution to connectivity.

Currently, few centralized or informal fishery systems have the resources and capacity to assess biodiversity impact beyond the boundary of the single fishery management area and the same can be said about most conservation networks. However, the situation is slowly improving, as shown in the Great Barrier Reef in Australia and Gulf of California.[18]

Notwithstanding, as for MPAs, the overall benefit from these interactions in a potentially large area, e.g. in the conservation network and the ecosystem, might not be easy to assess.

7.4.1.6   Biodiversity Impact Mitigation

According to the Biodiversity Impact Mitigation (BIM) framework supported by both IUCN[19] and the CBD[20] for development projects, the performance of a fishery-OECM may be taken to relate to the degree to which the area-based measures succeeded in successively: (i) avoiding collateral impact of fishing on the biodiversity attributes of concern; if not entirely possible because a zero impact is impossible or the feature is already impacted; (ii) reducing the current impact of fishing to the extent possible, and in any case below the level at which it would be considered a SAI or result in serious or irreversible harm; or (iii) rebuilding the biodiversity attributes of concern when they have been accidentally depleted below safe levels. The aim is to maintain the biodiversity values of concern at healthy levels (No Net Loss) when they have not suffered SAI or serious or irreversible harm, or otherwise to increase such values as required (Net Gain).

The Mitigation framework also foresees the use of biodiversity offsets, usually delivered at sites other than the impacted one, to compensate for any residual impacts after actions (i) to (iii) have been taken. However, the Decision refers to conservation outcomes being obtained *in situ* and UNCLOS specifies that all impacted resources must be maintained *in situ* or rebuilt at their safe level (Squires and Garcia, 2018). Consequently, in the current international ocean policy regime, offsetting may not be considered an option in OECM planning, management and international reporting of conservation coverage.

7.4.1.7   Additional properties

A comprehensive OECM performance also requires a check on the 'additional properties' (see sections 2.5 and 5.5.5) of OECMs, i.e. their representativeness, connectivity, complementarity with other protected areas around them, and their integration in broader networks (see Section 6.4), even in circumstances where these properties may not be expected to change much with time. The evolution of governance in terms of identification and participation of actors and equitable distribution of costs and benefits among them also needs to be checked. These properties were not determined in the OECM identification process but progress made in their regard during implementation would nonetheless help to improve the OECM performance.

7.4.2   Factors of effectiveness of fishery-OECMs

Few marine fishery-OECMs have been identified yet (at the end of 2024) and no thorough studies are available about factors that affect their effectiveness in producing broader biodiversity benefits. However, fishery-OECMs are 'special' ABMTs with a dual conservation purpose: (i) sustainable use of fishery resources, and (ii) improved conservation of biodiversity features of concern. They need to be effective with regard to both objectives: those aiming at the fishery sustainability, in order to be maintained as an ABFM, and those aiming at biodiversity conservation, to maintain OECM status. Consequently, the factors affecting effectiveness of fishery-OECMs are likely to be a mix of the factors already identified separately for ABFMs and MPAs in numerous publications.[21]

Despite differences in terminology and level of detail among the many references, there is a strong convergence overall on the factors that may affect 'effectiveness' across ABMTs used in fisheries (ABFMs), conservation (MPAs) and community-managed areas (including LMMAs, MMAs, etc.). These factors are certainly relevant also for fishery-OECMs, and interact in conditioning effectiveness. Most factors listed below spread over a continuous range of states or values, often split in discrete categories at specified benchmarks for various reasons and some examples follow. Factors may 'split' in two or a few states: species may be mobile or resident; habitats may be coastal or deep-sea, pelagic or benthic, degraded or intact; populations may be depleted, recovering or restored. They may take different values in a gradient (see Fig. 5.3), e.g. distance from coasts or human settlements; human population and density; fishing pressure. They may also be categorical, with the idea that while there may be variation within each category, there are genuine differences between them. Examples are categories of community size, of habitat types or of IUU fishing. Although other ways of categorizing factors of effectiveness may be possible, we suggest the following seven categories.

- *Physical parameters of the area*: Dimension (small to very large); ocean domain (e.g. coastal vs offshore; pelagic vs demersal; littoral vs deep-sea); type of substrate (e.g. mineral vs biogenic); geolocalization (clear vs non-existent); location (optimal vs suboptimal habitat).
- *Bioecological characteristics of the area*: Species concerned (resilient vs sensitive; highly mobile vs resident; widespread vs endemic; emblematic vs indifferent); population trends (recovering vs declining; stable vs variable).
- *Sources of pressures and threats* occurring or likely to occur in a foreseeable future such as low vs high levels of: fishing by all subsectors and IUU fishing; tourism; other economic pressures; climate change; land-based and local pollution; coastal degradation.

- *Human community parameters:* Urban vs rural settings; small communities vs large mega-cities; wealthy vs poor countries or jurisdictions; resilient vs vulnerable stakeholders; population increasing vs decreasing; cultural and social bonds active or decayed; developed or developing communities; availability vs lack of alternative livelihoods; effective vs corrupted leadership; existence or loss of ecological conscience, environmental ethics and cultural rituals; awareness of crises and opportunities, and of their potential costs and benefits; experience in collective action.
- *Policy*: Institutional frameworks co-ordinated or not; legal frameworks and institutional mandates adjusted for OECMs or not; focus on sustainable use and conservation, or siloed policies; financial support available for capacity building or not; economic and social incentives available or not.
- *Governance* is a central factor to effectiveness of ABMTs. In line with good governance principles (see sections 5.3.4 and 5.5.2, and Rice *et al.*, 2022), its expected properties include: (i) support for the identification of biodiversity, social and cultural features; (ii) inclusive processes; (iii) the self-regulation and consent of Indigenous Peoples; (iv) multiple levels of legitimate governance processes; (v) use of the best available science and local knowledge; (vi) clear legal status; (vii) management systems and plans in place or being developed; (viii) management integrated inside and outside the ABFM; (ix) monitoring systems in place; (x) assessment/feedback processes in place; (xi) cultural values and practices being respected; (xii) communities involved in monitoring and evaluation; (xiii) monitoring of social processes and benefits; (xiv) processes in place to ensure periodic reviews and recurrent evaluations; (xv) dependent community values respected; and (xvi) site vulnerability is assessed and considered.
- *Management* effectiveness may be related to: (i) implementation of the ecosystem approach; (ii) support from best scientific evidence and local knowledge; (iii) clarity of and support for applicable objectives; (iv) intended degree of protection; (v)

plans, rules and measures; (vi) elimination of destructive practices; (vii) effective and sustained monitoring and capacity-building programmes; (viii) diversification of management toolbox, including spatial and non-spatial measures; (ix) co-ordinated fisheries and conservation management strategies; (x) participative and deterrent enforcement; (xi) explicit connection of rights and responsibilities; (xii) graduated sanctions for non-compliance and inequity; (xiii) integration of measures taken inside and outside the OECM; (xiv) alternative income generation activities (AIGAS); (xv) auditing and accountability of authorities; and (xvi) conflict resolution mechanisms.

Many of these factors – which can increase effectiveness of fishery-OECMs or decrease it if absent – are likely to also be generally required for effective fisheries management. Most would be difficult to introduce in an otherwise weakly managed or unmanaged fishery system solely to attain and maintain fishery-OECM status, and areas weak in all the factors have probably not passed the quick screening identification stage. However, modern, well-funded and technically supported fishery management systems will have little difficulty in identifying and effectively using fishery-OECMs, should they wish to. The same is likely in community-managed fisheries with strong social connections, environmental stewardship and support from the State. Under other conditions, a proper use of fishery-OECMs will require significant capacity building in governance and management of the fishery sector.

The relative role of factors of effectiveness may be very case specific. For example, in the case of marine reserves, it has been shown that human population density and compliance had the strongest impact on fish biomass. However, the relationships between population density and the reserve effect were negative in the Caribbean, positive in the western Indian Ocean, and not detectable in The Philippines. No obvious reason for the differences could be found, implying that, in each case, different status on some 'other factors' had important roles in affecting fish biomass. Similarly, the relation between conservation performance and compliance was not always significant, and high levels of compliance were more related to

complex social interactions than simply to rules enforcement, stressing the importance of the social dimensions of ABMTs.

It has also often been suggested that using a combination of spatial and non-spatial management measures would improve the likelihood of meeting biological and socioeconomic fisheries goals (Hilborn *et al.*, 2004; O'Keefe *et al.*, 2014). Some factors, such as equity or compliance, are both factors and outcomes. An increase of either alone could certainly improve effectiveness but an increase in equity alone can promote an outcome of better compliance, which in turn is a factor that could potentially further improve effectiveness.

Altogether, the information in the many references given at the beginning of this section are complementary. There is broad agreement on what factors enable or limit the effectiveness of ABMTs as there is with MPAs. There is controversy, however, about cross-effectiveness, for example on the extent to which MPAs can be a major tool for improving fisheries management, economic optimization and food security, and conversely whether ABFMs granted OECM status may reliably contribute to conservation (e.g. as fishery-OECMs) and in the long term. In particular, there is some scepticism as to whether fisheries authorities in charge of fishery-OECMs will really continue to support the OECM measures promoting biodiversity conservation if the fishers complain that they are making the fisheries unsuccessful or too costly to operate.

Information available on effectiveness of MPA networks (potentially applicable to fishery-OECM networks) is limited (e.g. Green *et al.*, 2011; Grorud-Colvert *et al.*, 2014; Foster *et al.*, 2017; Meehan *et al.*, 2020). Such effectiveness would need to be measured against clear objectives defined by an adequate governance at the network level. It can be assumed that the factors affecting the effectiveness of an MPA network are closely related to that of its MPA components augmented by factors affecting network properties, such as ecological coherence, replication, ecological representation and connectivity (Cardoso-Andrade, 2022). It has been suggested that the importance of connectivity between MPAs was inversely proportional to the MPA size but that MPA network effectiveness is not affected by the distance between its MPA components (Kemp *et al.*, 2012) and conversely that

Table 7.1. Potential benefits and costs of OECMs.

Potential benefits	Potential costs
Enhanced existing ABFM biodiversity conservation co-benefits and, importantly, increased likelihood of meeting the 30% coverage target by 2030	Added management complexity and related costs in monitoring, assessment and enforcement
Further reduction or mitigation of fisheries' collateral impact on non-target species and habitats	Additional costs to the fishery sector if some existing fishing practices are excluded or displaced
Production of an incentive for better consideration of biodiversity outcomes in ABFM design	Raised interaction costs as the range of stakeholders increases with broader objectives
Improved connectivity of regional conservation networks of conservation measures	Risk for the fishery sector to tarnish its image if it fails to achieve or demonstrate expected outcomes
Strengthening of EAF implementation, facilitating, eco-labelling and related potential market benefits	Reduced co-operation of some fishery participants if their harvest-related objectives are being lowered
Improved image of fisheries with the public, consumers and civil society	Risk of losing part of the flexibility of fisheries management
More constructive collaboration between the conservation constituency and the fishery sector and its managers	Negative distributional impacts if some gain but others lose, especially if more marginalized players in the fishery lose as a result
Increased recognition and empowerment of local or shared management systems	Reduction of the overall effectiveness of fishery management if it leads to reduced compliance and greater enforcement problems

the smaller the MPAs of the network, the more important the role of connectivity (Thomas and Gillingham, 2015).

7.4.2.1   Costs and benefits

Any OECM – like any MPA or indeed any other management or conservation measure – will produce benefits and incur costs. Some potential costs and benefits are listed in Table 7.1. Note also that some of the costs and benefits to the sector or to biodiversity are non-monetary and hence not easily valued, but nevertheless can be important to consider in judging effectiveness.

The ratio of the benefits produced to the costs incurred ('benefit–cost ratio') provides an indication of the efficiency of the measure. However, it is often the case that the major impacts of a conserved area arise as a result of the distributional changes that occur – in which some gain but others lose (from the OECM). There is especially a risk that more marginalized players in the fishery, those with less bargaining power, are major losers as a result. Furthermore, the expected, perceived or measured 'benefits'

and 'costs' of recognizing or creating a fishery-OECM, and their distribution among stakeholders and right-sholders, influence actors' buy-in and compliance, and hence effectiveness.

Decision XIV/8 does not seem to directly address costs and benefits, or the trade-offs between them, but rather focuses on equity and costs and benefits sharing (thus directly relating to the distributional aspects above). The question is addressed in the Decision in relation to their monitoring, assessment, allocation and equitable sharing, and eventual alternative livelihoods and compensations (CBD, 2018: 18).

The process followed with OECMs is important to the relevant costs and benefits. When an existing ABFM is recognized as an OECM without any further expense, the costs of producing the biodiversity are already being incurred and absorbed in current management and fishing operations, raising few cost–benefit problems. However, if additional measures are needed to enhance the ABFM biodiversity benefits, the related additional cost and their distribution among actors will influence the decision, the actors' reactions and the outcomes. The question is particularly strategic in small islands developing states (SIDS) and least-developed countries, with limited budgets, monitoring and assessment capacity. In these environments, capacity building may be required incorporating local and Indigenous knowledge and local competences to improve effectiveness at lower costs.

7.4.3   How effective are fishery-OECMs likely to be?

All the above factors of effectiveness being considered relevant but to varying degrees case specific, and considering that it is too early to have any empirical assessment or fishery-OECMs' effectiveness, what information can be available about their likely effectiveness in the near future?

Even if a fishery OECM is well managed, its effectiveness in delivering desired biodiversity outcomes is likely to be lower than that of well-managed, strictly protected MPA reserves, and perhaps more comparable to that of multiple-use MPAs (IUCN Type VI) with important fisheries. Unfortunately, there have been few reviews of the current and probably variable future effectiveness of these 'partially protected areas', which also may include a proportion of 'paper parks'.

As shown in earlier chapters, fishery-OECMs have many similarities with ABFMs, often sharing many key factors of effectiveness. Therefore, to some extent the effectiveness of ABFMs in meeting their 'extended' fishery-related objectives may be a starting point for an early and cautious consideration on their potential performance as OECMs with broader biodiversity objectives. Much more information is available about ABFMs meeting fishery objectives than biodiversity objectives, but the relative importance of many of the factors may be similar for both types, even if the absolute magnitude of 'effectiveness' differs.

Modern fisheries management is strongly adaptive with assessments and reviews of outcomes scheduled on regular cycles. However, rarely are individual measures in a management plan evaluated separately for effectiveness because their effectiveness is entangled with that of other concomitant spatial and non-spatial measures also present and regulating the same fishery operations. Within fishery assessments, management effectiveness is judged primarily on the effect on biomass, productivity and catch level and composition, with strengths and weaknesses attributed to the *mix* of management measures (spatial and non-spatial) acting together, under the influence of external drivers. Despite these challenges of individual spatial measures being embedded in more comprehensive FMPs, numerous *ad hoc* reviews of assessments of fisheries with spatial measures in their FMPs have been made, and more are emerging.

In a thorough empirical assessment undertaken in the North Atlantic on ABFMs established for various purposes, ranging from stock sustainability to broader conservation purposes, STECF (2007) concluded that some positive outcomes were observed but no robust general conclusions could be reached because: (i) most of the measures considered had been established without formally stated objectives; (ii) the available information on stock status was too poorly spatially resolved to disentangle any closure effects from other identified causes of change; and (iii) it was not possible to determine to what extent the changes observed were related or not to the measures taken. The difficulty was confirmed in Rice *et al.* (2018) and Himes-Cornell *et al.* (2022) who also indicated that the effectiveness of ABFM subcategories (e.g. reserves, habitat or gear protection area, rotational closures, etc.) was highly variable, and dependent on many other factors in each case.

A systematic analysis of the literature available on 'fishery-ABMTs' (defined as ABMTs used in fisheries (Himes Cornell *et al.*, 2022) found that for conventional ABFMs, the results indicate that: (i) in 50% of the cases considered, broader positive biodiversity outcomes were mentioned in the related publication; (ii) in 67% of the cases, local relevant values were also addressed; (iii) and 15% of the ABFMs could be considered as potential fishery-OECMs based on a quick screening. In the case of community-managed areas, the results indicate that (i) 44% have produced broad positive biodiversity outcomes; (ii) 61% account for locally relevant values; and (iii) only 6% meet all OECM criteria and might therefore be considered as potential

OECMs containing traditional small-scale fisheries. It is also the case that 74% are already designated as MPAs, and thus cannot be fishery-OECMs, but the lessons about effectiveness of the measures in delivering positive biodiversity outcomes (nearly half the time), respecting local values (substantially more than half), but still not meeting all OECM criteria are still relevant.

Although these conclusions may refer mainly to effects on the target resources, the issues and the complexity of establishing clear cause–effect relationships would apply equally to the effects of fishery-OECMs and broader biodiversity features. The 'lowest hanging fruits' might be in ABFMs having conservation objectives as their primary objectives, such as strict fishery reserves or refuges, VMEs and other strictly protected essential fish habitats (Garcia et al., 2024).

The potential effectiveness of fishery-OECMs may also be considered through an *ex ante* assessment, based on expert opinion, using the OECM identification criteria. Such as assessment was conducted by Petza et al. (2019) for fisheries restricted areas (FRAs)[22] in the Aegean Sea, using proxies for the OECM criteria not yet finalized in the CBD process. A multicriteria decision analysis (MCDA) was fed by expert opinion to analyse 516 FRAs to (i) determine their likely level of contribution to marine biodiversity conservation in the Aegean Sea, (ii) categorize them according to the expectable level of effectiveness in biodiversity conservation, and (iii) provide scientific advice on which ones could be considered for subsequent analysis as potential OECMs. The MCDA characterized one FRA as 'extremely effective', one as 'very effective', 20 as 'effective', 147 as 'moderately effective', 264 as 'slightly effective' and 83 as 'ineffective'. If the first three categories were considered as having a high enough likelihood to produce the biodiversity benefits required for OECMs, they would represent just over 4% of the FRA population concerned.

More recently, in a systematic review of the literature on ABFMs covering the last 10 years, and considering only those for which the assessment was conclusive, Petza et al. (2023) concluded that, in relation to conventional fisheries management objectives, the social, economic and environmental outcomes were positive, mixed or negative respectively in 56%,

14 and 22% of the cases. However, the authors stressed that the literature scanned was very scarce in biodiversity effects.

7.5 Reporting through the Legitimate Authority

Decision 14/8 (Annex IV, Section 4) refers to 'reporting' on OECMs, asking Parties to:

a. Improve the frequency and accuracy of reports, maximizing use of existing mechanisms;

- enhance the reports' visibility and encourage their broad multidisciplinary analysis
- ensure that management is well informed to facilitate adaptive management
- build the capacity to report and analyse management effectiveness analyses
- build the political support for timely and effective reporting
- engage Indigenous Peoples and local communities in assessment and reporting
- develop and foster communities of practice.

Regular reporting on the performance of fishery management measures – whether area based or not – is one of the most important purposes of the fishery MER. These reports are needed by the legitimate authority primarily for adaptive management of the fishery but are also often a requirement to inform the government and justify the budget. For broader accountability and in accessible formats, these reports may also be made available to the public at large. Such reporting requires procedures, standards, schedules and formats established at the sector level (see Section 5.10.2).

At international level, the CBD Parties already have established processes for national reporting. In addition, Decision 14/8 (§6) encourages the relevant authorities to ... *submit data on OECMs to UNEP-WCMC for inclusion in the World Database on Protected Areas on OECMs (WDPA-OECMs) so that they can be taken into consideration when reporting on international conservation coverage targets.* The implications are discussed in Section 7.5.2.

7.5.1 Recurrent reporting

The suggested structure of a recurrent report to the legitimate authority is similar to that of the report on the identification process detailed in Section 5.6.2. The descriptions of the fishery and the OECM may be relatively stable across years, requiring little or no updating. The evolution of the biodiversity attributes of concern and their drivers is the central part of the recurrent performance report and will show whether attributes have been maintained (or improved) at the rate expected, or degraded. Negative outcomes (including improvement at a slower rate than expected) will require more comprehensive information, including the possible causes, e.g. changes in pressures and threats in and around the OECM (including climate change), poor management or weak compliance. Specific advice and 'options' that propose corrective action will be necessary for effective adaptive management, including what can be estimated about related costs and benefits and their distribution among stakeholders.

Considering the complexity of cause–effect relationships and feedback loops in complex social-ecological systems, the causal linkage between the changes observed in the OECM (or the fishery) and the various potential drivers of such change may not always be straightforward. Increasing challenges in determining causality make development of management options also increasingly challenging, in turn increasing both (i) the number of options for altering measures to deal with the unsatisfactory results of the periodic reviews, and (ii) the uncertainty of choices among the options. The adaptation of management measures may range from minor tweaking of the management regime, which can be done without affecting the status of the OECM, to its complete revision which might require a reassessment of the OECM status. The extent of the revision increases with the increase in negative or disappointing results However, implementing larger revisions to a management regime (which has usually evolved in small trial-and-error steps) is very often associated with higher uncertainties about their future consequences. That alone makes consensus on the particular actions harder to build, because the desired positive outcomes become less secure,

and even greater negative impacts of fishery performance become at least possible.

The frequency of the recurrent reports will be a local/national decision, based on the degree of urgency of the attributes' restoration, the attributes' expected rates of change, the evolution of the drivers of change, and the means available. In all cases, the timing of these reviews should be agreed when the MER is established. A less frequent reporting scheme may be adequate when multiannual management plans are implemented. Time lags can be expected between the implementation of measures in a fishery-OECM and the detection of many of their impacts in the ecosystem monitoring, justifying some time lag between adapting measures and evaluating their consequences. On the other hand, in cases of high uncertainty in choosing among management options and cases of particularly urgent OECM objectives (e.g. in relation to threatened species), a review of progress soon after the measures are revised may be requested, and frequent reviews thereafter.

The necessary scope of individual fishery-OECM reviews will depend on the nature of the objective and the threats and pressures addressed by the OECM. When a review of performance is requested soon after substantial revisions in the FMP, key indicators likely to reflect the success (or not) of the measures quickly should be identified when the revisions to the FMP are made. Reviews outside the agreed schedule may be needed in cases of significant change in management around the OECM, such as large changes in the size of the Total Allowable Catch (TAC) and allocations, or when the results of other research programmes may significantly affect the understanding of the basis on which the OECM was identified and managed, with the scope to review and report depending on the nature of the changes.

Reports should be made publicly available, with due considerations of individuals' confidentiality rights and the standards of FPIC.

As a separate document, or as a clearly separate section, the OECM MER system could report on its own implementation process and, for example, on (i) financial, human and technological resources mobilized; (ii) participatory processes; (iii) sampling schedule; (iv) new data collected; (v) assessment competences mobilized; (vi) extent to which the OECM MER implementation plan was respected; and (vii) lessons learned and improvements suggested, including an analysis of their costs and benefits (see next section). Similarly, the interaction, synergies or conflicts between the OECM MER and the fishery MER should be reviewed.

7.5.2 Reporting to UNEP-WCMC

Decision 14/8 (§5b) encourages State Parties to submit data on OECMs to the United Nations Environment Programme's World Conservation Monitoring Centre (UNEP-WCMC) for inclusion in the World Database on OECMs (WD-OECM). A State doing so can have its OECMs area accounted as part of its contribution to international biodiversity conservation coverage targets (e.g. in the GBF or the SDGs). However, current CBD Decisions do not *obligate* States to report their OECMs, should they choose not to. Both consistency and transparency would be incentives for each legitimate authority to adopt the practice of either all nationally recognized OECMs being consistently registered in the WCMC OECM database, or another method for public communication of the OECMs should be established. Case-by-case choices to register OECMs with the WCMC or not would be confusing to both the public and many agencies and interest groups following the relevant Target.

Incentives for reporting to the WCMC include: (i) assistance provided by the WCMC for coherent recurrent reporting; (2) having the information included in a unique inventory of conservation areas of different types and origin, with regrouped and checked metadata, accessible on maps, providing also *de facto* a 'national repository' when this is not locally available; (iii) capacity-building opportunities offered by UNEP-WCMC; (iv) contribution to data sources already used for research and spatial management; (v) reuse of geolocated data at regional level, e.g. in RFMOs/As or LMEs (regional observatories) (modified from UNEP-WCMC, 2019).

The elements needed for such initial reporting are indicated in the WCMC OECM reporting manual (UNEP-WCMC, 2019).[23] The first obvious overriding requirement is that the OECM must meet the CBD definition of an OECM, which requires evidence that the OECM

has been evaluated relative to the CBD criteria and they are met. The 'minimum attributes' that are required for registration of an OECM (hereafter underlined) and the 'complete attributes' which the data providers are also encouraged to submit are explained in detail in the manual, e.g. (i) area category (MPA or OECM); (ii) name of the OECM in local or native language(s) and English; (iii) designation name (e.g. refugia, closed area, fishery reserve); (iv) designation type, e.g. national, regional, international, transboundary); (v) marine area covered by the OECM; (vi) total area covered (for OECMs including both marine and terrestrial areas); (vii) status (e.g. proposed, established, candidate); (viii) are no-take areas included; (ix) status year (when the OECM was proposed, established, etc.); (x) type of governance and evidence of consent by all legitimate authorities; (xi) type of ownership; (xii) management authority; (xiii) management plan; (xiv) supplementary info, e.g. full assessment report showing how criteria are met; (xv) conservation objectives; (xvi) location codes: ISO alpha-3 codes.

Most of these attributes will be part of the national identification reports (see Section 5.6.2), and if the WCMC attributes are kept in mind while the national report is developed, that report can be used with very minor modifications for the report to the WCMC. Additional (optional) information can also be submitted. The sources of any uploaded data must be provided. The intellectual property rights must be specified in the Data Contributor Agreement which must be signed and state *inter alia* how the data provided will be used, and that redistribution or use of the data by third parties will be subject to the WD-PA Terms of Use.

The UNEP-WCMC manual provides also information on: (i) who can report to WCMC; (ii) the intellectual property rights on the data; (iii) the rules of access and use of the data; (iv) the process of uploading into the OECM database and related data standards; (v) the data verification processes (in case the data are not provided by a government); (vi) revisions of the data (at the data provider initiative or on call from the WCMC, e.g. every 5 years).

The reporting of OECMs by RFMOs/As has not been specifically considered in Decision 14/8. However, nothing impedes States Parties to the CBD seeking to implement their commitments (e.g. on OECMs) under another convention to which they are also Parties, such as a RFMO/A, and RFMO input was welcomed at CBD COP 14 where Decision 14/8 was adopted. Indeed, particularly since UNGA Resolution 61/105, all RFMOs have identified the VMEs within their jurisdiction, and many RFMOs/As already have adopted measures, including spatial measures, to manage the impact of fisheries on biodiversity in their areas of jurisdiction. RFMOs/As usually make their assessments and management performance reports available on their websites. Consequently, it has been suggested (in Garcia *et al.*, 2022) that, with agreement of their Parties, RFMOs/As could report on OECMs to the WCMC database, as the most competent data providers, on behalf of their member governments. Indeed, the WCMC manual indicates that data providers include secretariats of international conventions and regional entities.

Since the adoption of Decision 14/8, and particularly following the ICES-FEG Workshop on Testing OECM Practices and Strategies (WKTOPS) (ICES, 2021), a number of RFMOs, including NAFO, NEAFC and GFCM, have started to work towards identification of fishery-OECMs in their areas of competence. After some interaction with the WCMC Secretariat aiming at clarifying the process, and particularly the RFMO's Parties decision on OECM identification and the process of evaluation of their effectiveness, some OECMs established by NEAFC and NAFO were registered in April 2025. As a by-product of this process, some additional guidance on fishery-OECMs may be prepared in the months to come about the registration process of OECMs by RFMOs and on the evaluation of their effectiveness, based on the long experience of RFMOs in evaluating performance in their fishery management systems, and taking due consideration of the specific broader biodiversity requirements (see Boxes 2.1, 4.1 and 4.2).

7.6 Archiving and Communication

For reasons of long-term data safety, effectiveness of the assessment process and institutional memory, and to facilitate data retrieval, assessment updating and adaptive management, it is also valuable to archive all the information and advice leading to decisions regarding OECMs in a national repository (or, if not available, a sectoral repository). The data collected and information generated by the assessments represent a significant cost as well as an asset of significant economic value for the adaptive management system. As such, they need to be preserved (archived and maintained) and communicated broadly to all interested stakeholders, including auditors.

7.6.1 Archiving

The reports and all related numerical or narrative data and information, including local knowledge, should be safely stored in an information management system established in an accessible archive of an appropriate governmental agency, carefully considering the information standards (e.g. formats, software, languages) and rules, to ensure proper data input and consistency checks, workflows, data access and exchange protocols (confidentiality) and integrity of the databases. This is important for historical and institutional memory, retrospective reviews and long-term performance appraisal, consistency in monitoring and evaluation, adaptive management, etc. Because of the degree of flexibility granted by the CBD Decision for the assessment and management approaches used, systematic archiving allows maintenance of the 'pedigree' of each OECM (e.g. date of creation; legitimate authority responsible for the creation; data available; time period covered; methods used; measures applied in the OECM; results obtained, etc.) which allows an objective opinion on the relative robustness of the OECM and an informative track of its evolution.

As the OECMs identified in fisheries will usually have objectives for both sustainability of the fishery and specifically regarding biodiversity outcomes, integrated in the FMP, the OECM information will conveniently be archived in the same information repository as other related and interconnected measures taken in the sector. However, as they also contribute to conservation and may have strong cross-sectoral dimensions, they may also need to be archived with other ministries and/or at other levels of governance, reflecting national policy and practice.

The task of keeping all national repositories synchronized is not to be underestimated but is not unique to OECMs and their supporting information. In additional to national efforts to maintain consistency among its OECM archives, the WCMC database represents *de facto* a unique inventory of conservation areas of different types and origin, with regrouped and checked metadata, accessible on maps as a proxy national observatory. The national (and WCMC) registries each serve as institutional memory and can be used for recurrent monitoring and evaluation of long-term performance assessment.

7.6.2 Communication to a diverse audience

The practice of broadly communicating on governance decisions and performance is part of equitable governance good practices. The results of the OECM identification process, their performance and corrective decisions,should officially be communicated by the legitimate authority to the auditors as well as all fisheries managers and fishers and other stakeholders such as the scientific community and conservation and fisheries advocacy groups. Where initially some of these targeted communication audiences may not be familiar with OECMs, appropriate background and context information should also be made available, ideally with in-person meetings. In particular, feedback information will be appreciated by all those who have contributed time, information and competence to the process, using adequate communication means (e.g. governmental channels, social media, beach radios and TV news). This should be done also in local languages where appropriate, taking into account diverse audiences in fisheries, conservation and other sectors. This process will be facilitated if, consistent with principles of equitable governance, a high level of active participation has been provided throughout the entire process. It may require specific efforts to tailor the information communicated to different types of recipients, with different levels of formal education, and in local languages.

7.7 Auditing

Auditing is part of the general task of evaluating performance of any programme, particularly when using public funding or funds from a foundation or other donor sources. Auditing is not explicitly addressed in Decision 14/8 and is therefore not formally required for OECM identification and management. However, auditing would be important to reassure the fisheries and conservation stakeholders of the quality of the OECM management and of its alleged outcomes, and it could be undertaken at national level if decided by the legitimate authority or the State, for oversight and accountability. Auditing may also be undertaken by an accredited third party. In such a case, MER reports should systematically be submitted also to the auditors.

Ideally, for practical reasons, OECM auditing should be undertaken as part of the auditing of the broader fishery management of the fishery/sector within which the OECM(s) operate, unless decided otherwise, e.g. because of the urgency required by some threatened biodiversity attribute. In the context of OECM performance, an audit would seek to ensure that: (i) the indicators of performance of the OECM in relation to the biodiversity attributes of concern fairly reflect the performance of the OECM as required in the CBD Decision 14/8 principles and criteria; and (ii) the MER programme is conducted in an economical, efficient and effective manner (adapted from INTOSAI WGEA, 2007).

The action needed would include the following.

• Defining the auditing protocol. This may be done by the auditors with the legitimate authority and in collaboration with the MER authority, so that the MER is well informed and can collect and archive all the necessary information.

• Auditing the performance of the OECM(s) against OECM principles and criteria, possibly in connection with the audit of the FMP itself.

• Auditing the information content, relative to the OECM principles and criteria, of the indicators, models and analytical methods used in the initial evaluation of the ABFM for OECM status, and subsequently for the performance evaluations, and the costs to maintain the indicators and models.

- Auditing the functioning of the OECM MER itself to certify the wise use of funds and resources.
- Communicating the non-confidential conclusions of the audit to fishery and conservation managers and to the public through available communication means.

There is no detailed guidance yet on how to conduct such an audit of an OECM but there is broad guidance on auditing and particularly on auditing biodiversity (e.g. INTOSAI WGEA, 2007), environmental auditing (INTOSAI WGEA, 2007a) and auditing in the perspective of Sustainable Development Goals (SDGs) (INTOSAI WGEA, 2019).

7.8   Revision of the OECM Status

Minor oscillations of the components of OECM performance from year to year are to be expected, due to natural oscillations in the ecosystem as well as estimation errors or minor changes in fishing operations. Appropriate corrective action falls within the remit of 'ordinary' adaptive management of the OECM (see Chapter 6) and will not be discussed here.

The OECM MER may instead identify more important shortcomings in the OECM outcomes, related *inter alia* to non-optimal location or size of the area, important changes in the local environment, in local socioeconomic conditions and compliance, or due to climate change, for which simple tweaking of the OECM management might not be sufficient, and 'extraordinary' measures would logically be recommended. In this case, the legitimate authority could logically consider a major revision of the OECM characteristics and regulations inside the area and around it, to improve performance. In the probably rare case in which no improvement was considered possible, cost-effective or otherwise worthwhile, the ABFM should lose its OECM status. Related actions at the national and subnational level would depend on internal processes and standards, but would be expected to typically result in the OECM being delisted from the national archive. Consequently, the area would cease to be reported under global

biodiversity targets and the decision would be reported in the WCMC database.[24]

The exact processes for making such important decisions have not been spelled out anywhere yet, and are currently at the discretion of the legitimate authority. The CBD has adopted a detailed process for how the boundaries of EBSAs can be adjusted (CBD COP Decision XVI/16, 2024), but comparable formal guidance on changes of OECM descriptions has not been developed. A formal guidance from the CBD would be useful on the frequency with which OECM status should be reviewed for effectiveness and provision of benefits; on how existing descriptions, including boundaries, should be revised; and on how significant a change must be before a new OECM identification process is required. Considering that a modest change in location may have more importance for the status of small OECMs with boundaries defined by distinct habitat features than for larger OECMs with less distinct ecological boundaries, it is likely that a case-by-case approach to this issue will again be necessary.

As a default, if they are available, the same bodies and processes involved in the initial identification of the OECM (described in Section 5.5), or alternative ones preferred by the legitimate authority, would be activated to reassess the OECM, considering the new information provided by the MER and its implications for the OECM status, with the thoroughness applied in the identification phase. Depending on the nature of the shortcomings, only some of the identification steps might need to be fully replicated with inclusion of the new information.

Ideally, it would also be useful to determine the type of triggers and threshold values in OECM properties that would lead to recommending a major revision or deletion from reporting, integrating them in management decision rules. This would improve management reactivity. However, there is presently too little experience with OECMs to determine such values.

Actions possibly undertaken prior to and for revisions are described below.

Prior to the revision, it would be useful to determine their periodicity to ensure that the information is regularly updated. Based on scientific and other evidence, the legitimate authority may decide whether revisions: (i) should occur by default with their own periodicity, although special conditions (to be listed) could lead the legitimate authority to call for an earlier review and possible revision (e.g. an event like a major tsunami or oil spill that could alter the ecosystem) or postpone the review (e.g. as was necessary for many reviews during the Covid pandemic); (ii) are synchronized with the process established to revise the FMP, whether annual or multiannual; (iii) are synchronized with the reporting schedule established for the reporting to the WCMC (see Section 7.5.2); or (iv) are triggered by negative OECM MER conclusions regarding performance.

When undertaking a revision, the following may be considered.

- Following the assessment process of identification (see Chapter 5), retrieve the information used for identification and that produced by the MER. Identify the steps that might have to be revised. The factors for an effective revision are similar to those for identification (e.g. research capacity, collaborations, available data).
- Avoid 'overreaction' and take into account expected 'natural' variations in system components as well as uncertainties in assessment and management before modifying the management regime. With time, the 'normal variance' of performance may

be appraised and explicit tolerances for deviations from the expected trajectories could be set. However, at the beginning, there is likely to be a weak basis to decide whether an observed change is a 'signal' of concern, signalling inadequate performance of the OECM, or a 'noise' characteristic of the system, to be noted but not calling for a regime change. High variability in the 'properties' of an area would justify caution in its identification as an OECM, but if the area is found to meet the criteria adequately, that degree of variability should be considered acceptable in performance as well.

- If needed, temporarily suspend the OECM from the WCMC database,[25] while withholding a more permanent decision on its total removal until there is greater confidence that the unfavourable status and outcomes of the area are likely to persist. If the MER reports are timely and of good quality and if management responses to them are swift and effective, the corresponding revisions of specific measures or expectations could be fast enough to avoid a suspension.
- Report as appropriate about the revision to the CBD and in the OECM database handled by the WCMC.

Notes

[1] In practice, the interactions between the different measures, in and out of the ABFM, the predator–prey relationships and trophic cascades, the environmental and socio-economic forces, etc. are such that the effect of a single spatial measure on the fishery can often only be 'demonstrated' by modelling, under a demanding set of assumptions.

[2] Such as economic returns, employment created, wellbeing of individuals and communities, social sustainability, human rights and the triple bottom-line balance between ecological, economic and social dimensions of the fishery and its management.

[3] See also California, MPA monitoring action plan: available at www.wildlife.ca.gov/Conservation/Marine/MPAs/Management/Monitoring/Action-Plan: MPA Watch at www.mpawatch.org/ (accessed 28 July 2025).

[4] For example, historical information, surrounding ecosystems; geolocation; sources of data and information; stakeholders and potential collaborations; species; habitats; biodiversity features of concern; assessment methods, fishing activities; management measures in and around the area; and bibliography.

[5] The performance of the OECM in relation to the fishery and target resources sustainability (usually its primary objective) is of great relevance for the fishery sustainability and Aichi Target 6 and successor targets and for SDGs. Abundant guidance is available for the purpose and this subject is not addressed here.

[6] Secondary important outputs are produced in the process, such as a data and information management system to maintain records in the long term.

[7] The concept of 'key performance measure' is not foreseen in Decision 14/8 but would reflect the degree of 'flexibility' that the Decision provides for States in implementing the Decision.

[8] The use of the BIM hierarchy is recommended in CBD Decision 14/8, Annex IV, C, 5e).

[9] However, reference is hard, if not impossible, to apply for numerous poorly known species and for threatened habitats.

[10] If the same species were affected negatively inside the OECM, by pressures from other sectors, e.g. by collision with tankers, and the expected biodiversity benefits from fisheries measures could not be ensured, the OECM ought to be delisted or, better, a cross-sectoral agreement could be sought (with the lead or support of the State as necessary) to eliminate, reduce or mitigate the external impact. In case of climatic unfavourable events or trends (e.g. in the case of reduced productivity, coral bleaching), efforts should be made to proportionally reduce the fishing pressure in the OECM and possibly in the fishery itself.

[11] The performance of the OECM in relation to the fishery and target resources sustainability is of great relevance for the fishery sustainability and Aichi Target 6 and successor targets and for SDGs. Abundant guidance is available for the purpose and this subject is not addressed here. See ,for example, Hockings (1998), Pomeroy et al. (2004, 2005), Fancy et al. (2008), Field et al. (2004, 2005, 2007), FAO (2003, 2009a, b), Cochrane and Garcia (2009), and Lindenmayer and Likens (2010).

[12] The fishery sector may not have the capacity to face the cost and have the competence to assess trends in threats emanating from other sectors, but the MER must have some ability to at least inform itself on such threats, as far as they may be foreseeable, and report on what is known about them, based on external sources.

[13] The difference between the expected and actual outcomes of the fishery OECMs affect the performance of the OECM measure (its location, technical measures, etc.) and the management system, as described in Section 7.3.1.

[14] Because of the numerous interactions between the OECM 'ecosystem' and the broader ecosystem around it and in which the fishery operates.

[15] This section draws heavily from Garcia et al. (2024) on ABMTs and fisheries.

[16] However, that effectiveness in relation to the strict 'fishery-related' objectives, relevant for other GBF targets, is not taken into account when assessing the effectiveness of the OECM, even though the fishery resources are an integral part of the biodiversity of the fishing ground.

[17] For example, through growth and reproduction of individuals or life stages of the biodiversity features of concern exported by the OECM.

[18] https://wildlife.ca.gov/Conservation/Marine/MPAs (accessed 29 July 2025).

[19] https://iucn.org/our-work/topic/biodiversity-net-gain (accessed 29 July 2025).

[20] www.google.com/search?q=Mitigation%20hierarchy%20site%3Acbd.int (accessed 29 July 2025).

[21] For example, Pollnac (1994), Crawford et al. (2000), Salm et al. (2000), WRI (2000), Hilborn et al. (2004), Pomeroy et al. (2005), Kaplan (2009), Ehler and Douvere (2009), Cox et al. (2010), Claudet et al. (2011), Green et al. (2011), Kearney and Farebrother (2012), Mesnildrey et al. (2013), Devillers et al. (2014), O'Keefe et al. (2014), Edgar et al. (2014), Grorud-Colvert et al. (2014), Rocliffe (2015), Foster et al. (2017), Ban et al. (2017), Gill et al. (2017), Rice et al. (2018), Vivacqua (2018), Meehan et al. (2020), Cardoso-Andrade (2022), Garcia et al. (2024).

[22] Fisheries restricted areas (FRAs) are geographically defined areas in which all or certain fishing activities are temporarily or permanently banned or restricted by fisheries, environmental, archaeological or maritime legislation at the national, European or international level (e.g. by the GFCM).

[23] www.protectedplanet.net/c/wdpa-manual/wdpa-manual-v16 (accessed 29 July 2025).

[24] UNEP-WCMC already operates a 'take-down' policy, allowing withdrawal of all or a portion of the data from the database under various circumstances such as breach of copyright, confidentiality, defamation or libel.

[25] The WCMC manual does not refer to this option. Therefore, its feasibility should be checked with the WCMC Secretariat which is dedicated to support CBD parties in their effort.

References

Ablan, M.C.A., McManus, J.W. and Viswanathan, K. (2004) Indicators for management of coral reefs and their applications to marine protected areas. *NAGA* 27(1): 31–39.

Addison, P.F.E., Carbone, G. and McCormick, N. (2018) *The Development and Use of Biodiversity Indicators in Business: An Overview.* IUCN, Gland, Switzerland.

Afflerbach, J.C., Lester, S.E., Dougherty, D.T. and Poonc, S.E. (2014) A global survey of 'TURF-reserves', Territorial Use Rights for Fisheries coupled with marine reserves. *Global Ecology and Conservation* 2: 97–106.

Aften, T. and Fuller, S. (2019) A Technical Review of Canada's Other Effective Area- Based Conservation measures: Alignment with DFO Guidance, IUCN WCPA Guidance and CBD SBSTTA Guidance. Available at: https://davidsuzuki.org/science-learning-centre-article/a-technical-review-of-canadas-other-effective-area-based-conservation-measures-alignment-with-dfo-guidance-iucn-wcpa-guidance-and-cbd-sbstta-guidance/ (accessed 20 July 2026).

Alban, F., Boncoeur, J. and Roncin, N. (2011) Socioeconomy – assessing the impact of marine protected areas on society's well-being: an economic perspective. In: Claudet, J. (ed.) *Marine Protected Areas. A Multidisciplinary Approach.* Cambridge University Press, Cambridge, UK, pp. 226–246.

Allen, P.M. and McGlade, J.M. (1987) Modelling complex human systems: a fisheries example. *European Journal of Operational Research* 30: 147–167.

Anaya, J. (1996) *Indigenous Peoples in International Law.* Oxford University Press, New York, USA.

Anderson, J.L., Anderson, C.M., Chu, J., Meredith, J., Asche, F. *et al.* (2015) The fishery performance indicators: a management tool for triple bottom line outcomes. *PLoS ONE* 10(5): e0122809.

Atkinson, S., Esters, N., Farmer, G., Lawrence, K. and McGilvray, F. (2011) *The Seascapes Guidebook: How to Select, Develop and Implement Seascapes.* Conservation International, Arlington, USA.

Bagri, A. and Vorhies, F. (1997) Biodiversity impact assessment. Draft discussion paper for SBSTTA3, Montreal, Canada, 1-5 September.

Bamford, S. (2024) On being 'natural' in the rainforest marketplace: science, capitalism and the commodification of biodiversity. *Social Analysis* 46: 1–16.

Ban, N.C., Davies, T.A., Aguilera, S.E., Brooks, C., Cox, M. *et al.* (2017) Social and ecological effectiveness of large marine protected areas. *Global Environmental Change* 43: 82–91.

Barange, M., Bahri, T., Beveridge, M.C.M., Cochrane, K.L., Funge-Smith, S. and Poulain, F. (2018) *Impacts of Climate Change on Fisheries and Aquaculture. Synthesis of Current Knowledge, Adaptation and Mitigation Options.* FAO, Rome, Italy.

Bates, A.E., Cooke, R.S.C., Duncan, M.I., Edgar, G.J., Bruno, J.F. *et al.* (2019) Climate resilience in marine protected areas and the 'Protection paradox'. *Biological Conservation* 236: 305–314.

Bayley, D.T.I. and Mogg, A.O.M. (2019) New advances in benthic monitoring technology and methodology. In: Sheppard, C. (ed.) *World Seas: An Environmental Evaluation*, 2nd edn. Academic Press, New York, USA, pp. 121–132

Belfiore, S., Cicin-Sain, B.,and Ehler, C. (eds) (2004) *Incorporating Marine Protected Areas into Integrated Coastal and Ocean Management: Principles and Guidelines*. IUCN, Gland, Switzerland.

Berkes, F. (1999) *Traditional Ecological Knowledge and Resources Management*. Taylor and Francis, London, UK.

Bermejo, R. and Bermejo, R. (2014) The commodification of nature and its consequences. In: Bermejo, R. (ed.) *Handbook for a Sustainable Economy*. Springer, Cham, Switzerland, pp. 19–33.

Biggs, C.R., Yeager, L.A., Bolster, G.G., Bonsell, C., Dichiera, A.M. *et al.* (2020) Does functional redundancy affect ecological stability and resilience? A review and meta-analysis. *Ecosphere* 11(7): e03184.

Biodiversity Indicators Partnership (2010) *Biodiversity Indicators and the 2010 Target: Experiences and Lessons Learnt from the 2010 Biodiversity Indicators Partnership*. Secretariat of the Convention on Biological Diversity, Montreal, Canada.

Biodiversity Indicators Partnership (2011) *Guidance for National Biodiversity Indicator Development and Use.* UNEP World Conservation Monitoring Centre, Cambridge, UK.

BIONET and IUCN (1997) *Exploring Biodiversity Indicators and Targets Under the Convention on Biological Diversity*. Report of the Sixth Global Biodiversity Forum. Biodiversity Action Network (BIONET), Washington, DC, USA and IUCN, Gland, Switzerland.

Bloor, I., Dignan, S., Emmerson, J., Beard, D., Gell, F.E. *et al.* (2021) Boom not bust: cooperative management as a mechanism for improving the commercial efficiency and environmental outcomes of regional scallop fisheries. *Marine Policy* 132: 104649.

Borrini-Feyerabend, G. and Hill, R. (2015) Governance for the conservation of nature. In: Worboys, G.I., Lockwood, M., Kothari, A., Feary, S. and Pulsford, I. (eds),*Protected Area Governance and Management*. ANU Press, Canberra, Australia, pp. 169–206.

Borrini-Feyerabend, G., Dudley, N., Jaeger, T., Lassen, B., Pathak Broome, N. *et al.* (2014) *Gouvernance des Aires Protégées: de la Compréhension à L'action*. IUCN, Gland, Switzerland.

Bourgeron, P.S., Humphries, H.C. and Jensen, M.E. (2001) Representativeness assessments. In: Jensen, M.E. and Bourgeron, P.S. (eds) *A Guidebook for Integrated Ecological Assessments*. Springer, New York, USA, pp. 292–306.

Bruno, J.F., Bates, A.E., Cacciapaglia, C., Pike, E.C., Amstrup, S.C. *et al.* (2018) Climate change threatens the world's marine protected areas. *Nature Climate Change* 8(6): 499–503.

Butler, J.R.A., Tawake, A., Skewes, T., Tawake, L. and McGrath, V. (2012) Integrating traditional ecological knowledge and fisheries management in the Torres Strait, Australia: the catalytic role of turtles and dugong as cultural keystone species. *Ecology and Society* 17(4).

Butsic, V., Lewis, D., Radeloff, V., Baumann, M. and Kuemmerle, T. (2017) Quasi-experimental methods enable stronger inferences from observational data in ecology. *Basic and Applied Ecology* 19: 1–10.

Caddy, J.F. (2002) Limit reference points, traffic lights, and holistic approaches to fisheries management with minimal stock assessment input. *Fisheries Research* 56(2): 133–137.

Cadrin, S.X. and Dickey-Collas, M. (2015) Stock assessment methods for sustainable fisheries: introduction. *ICES Journal of Marine Science* 72(1): 1–6.

Canales, C.M., Hurtado, C. and Techeira, C. (2018) Implementing a model for data-poor fisheries based on steepness of the stock-recruitment relationship, natural mortality and local perception of population depletion. The case of the kelp *Lessonia berteroana* on coasts of north-central Chile. *Fisheries Research* 198: 31–42.

Cardoso-Andrade, M., Queiroga, H., Sousa, R. M., Belackova, A., Bentes, L. *et al.* (2022) Setting performance indicators for coastal marine protected areas: an expert-based methodology. *Frontiers in Marine Science* 9: 1–13.

CBD (2004) *The Ecosystem Approach (CBD Guidelines)*. Secretariat of the Convention on Biological Diversity, Montreal, Canada. https://www.cbd.int/doc/publications/ea-text-en.pdf

CBD (2005) *Indicators for Assessing Progress Towards, and Communicating, the 2010 Target at the Global Level.* United Nations Environment Programme, Nairobi, Kenya.

CBD (2006) Voluntary Guidelines on Biodiversity-Inclusive Impact Assessment (COP Decision VIII/28). Available at: www.cbd.int/doc/decisions/cop-08/cop-08-dec-28-en.pdf (accessed 29 July 2025).

CBD (2010) The Strategic Plan for Biodiversity and the Aichi Biodiversity Targets. Decision X2 taken at CBD COP 10, Nagoya, Japan, 18–29 October. Document UNEP/CBD/COP/DEC/10/2: 13.

CBD (2012) Marine and Coastal Biodiversity: Revised Voluntary Guidelines for the Consideration of Biodiversity in Environmental Impact Assessments and Strategic Environmental Assessments in Marine and Coastal Areas. Document UNEP/CBD/COP/11/23 submitted to the COP Eleventh Meeting, Hyderabad, India. Available at: www.cbd.int/doc/meetings/cop/cop-11/official/cop-11-23-en.pdf (accessed 29 July 2025).

CBD (2018) Decision 14/8. Protected Areas and Other Effective Area-Based Conservation Measures. Adopted by the 14th Conference of the Parties of the Convention on Biological Diversity, Sharm El-Sheikh, Egypt, 17–29 November. Document CBD/COP/DEC/14/8. Available at: www.cbd.int/doc/decisions/cop-14/cop-14-dec-08-en.pdf (accessed 29 July 2025).

CBD (2023a) Report of the Sustainable Ocean Initiative (SOI) national workshop for Jamaica on other effective area-based conservation measures (OECMs) in the marine fishery sector. Kingston, Jamaica, 17–19 May. Available at: www.cbd.int/meetings/SOI-WS-2023-02 (accessed 29 July 2025).

CBD (2023b) Report of the Technical Expert Workshop to Review Modalities for Modifying the Descriptions of Ecologically or Biologically Significant Marine Areas and Describing New Areas. Document CBD/EBSA/EM/2023/1/3.

CBD (2023c) Report of the legal expert workshop to review modalities for modifying the descriptions of ecologically or biologically significant marine areas and describing new areas. Document CBD/EBSA/EM/2023/2/3.

CBD (2023d) Report of the Conference of the Parties to the Convention on Biological on the second part of its 15th meeting. Montreal, Canada, 7–19 December and Nairobi, 19–20 October. Document CBD/COP/15/17.

CCFAM (2017) Report on Canada's Network of Marine Protected Areas. Canadian Council of Fisheries and Aquaculture Ministers (CCFAM), Ottawa, Canada. Available at: www.dfo-mpo.gc.ca/oceans/publications/oeabcm-amcepz/index-eng.html (accessed 29 July 2025).

Chambers, M., Wyborn, C., Ryan, M.E., Reid, R.S., Riechers, M. *et al.* (2021) Six modes of co-production for sustainability. *Nature Sustainability* 4: 983–996.

Charles, A. (ed.) (2021) *Communities, Conservation and Livelihoods*. IUCN/Community Conservation Research Network, Gland, Switzerland.

Charles, A. (2023) Fishery knowledge. In: *Sustainable Fishery Systems*, 2nd edn. Wiley, Chichester, UK.

Charles, A. and Wilson, L. (2009) Human dimensions of marine protected areas. *ICES Journal of Marine Science* 66: 6–15.

Charles, A., Macnaughton, A. and Hicks, S. (2024) *Environmental Stewardship by Small-scale Fisheries*. FAO, Rome, Italy.

Charles, A.T. (2001) *Sustainable Fishery Systems*. Blackwell Science, Oxford.

Chase, J.M. and Leibold, M.A. (2003) *Ecological Niches: Linking Classical and Contemporary Approaches*. University of Chicago Press, Chicago, USA.

Chesson, J. (2013) Sustainable development. Connecting practice with theory. *Journal of Environmental Assessment Policy and Management* 15(1): 27.

Chesson, J. and Clayton, H. (1998) *A Framework for Assessing Fisheries with Respect to Ecologically Sustainable Development*. Bureau of Rural Sciences, Canberra, Australia.

Christie, P., Bennett, N., Gray, N., Wilhelm, A., Lewis, N. *et al.* (2017) Why people matter in ocean governance: incorporating human dimensions into large-scale marine protected areas. *Marine Policy* 84: 273–284.

Chuenpagdee, R., Morgan, L.E., Maxwell, S.M., Norse, E.A. and Pauly, D. (2003) Shifting gears: assessing collateral impacts of fishing methods in US waters. *Frontiers in Ecology and the Environment* 1(10): 517–524.

Claudet, J., Guidetti, P., Mouillot, D., Shears, N.T. and Micheli, F. (2011) Ecological effects of marine protected areas: conservation, restoration and functioning: In: Claudet, L. (ed.) *Marine Protected Areas. A Multidisciplinary Approach*. Cambridge University Press, Cambridge, pp. 37–71.

Cochrane, K.L. and Garcia, S.M. (2009) *A Fishery Managers' Guidebook*, 2nd edn. Wiley-Blackwell, Chichester, UK and FAO, Rome, Italy.

Cohen-Shacham, E., Walters, G., Janzen, C. and Maginnis, S. (eds) (2016) *Nature-based Solutions to Address Global Societal Challenges*. IUCN,Gland, Switzerland.

Coll, M., Carreras, M., Ciércoles, C., Cornax, M.-J., Gorelli, G. *et al.* (2014) Assessing fishing and marine biodiversity changes using fishers' perceptions: the Spanish Mediterranean and Gulf of Cadiz Case Study. *PLoS ONE* 9(1): e85670.

Collie, J.S., Botsford, L.W., Hastings, A., Kaplan, I.C., Largier, J.L. *et al.* (2016) Ecosystem models for fisheries management: finding the sweet spot. *Fish and Fisheries* 17(1): 101–125.

Constable, A.J. (2011) Lessons from CCAMLR on the implementation of the ecosystem approach to managing fisheries. *Fish and Fisheries* 12: 138–151.

Costanza, R. (1999) The ecological, economic, and social importance of the oceans. *Ecological Economics* 31(2), 199–213.

Costanza, R., d'Arge, R., de Groot, R., Farber, S., Grasso, M. *et al.* (1997) The value of the world's ecosystem services and natural capital. *Nature* 387: 253–260.

Couture, E. and Rideout, R. (2014) Standardizing the Traffic Light Approach for Reporting on Convention Objectives. NAFO Report SCR Doc. 14. Available at: https://www.nafo.int/Portals/0/PDFs/sc/2014/scr14-045.pdf (accesssed 29 July 2025).

Cowie, W., Al Dhaheri, S., Al Hashmi, A., Solis-Rivera, V., Baigun, C. *et al.* (2020) IUCN Guidelines for Gathering of Fishers' Knowledge for Policy Development and Applied Use. IUCN/Environment Agency, Gland/Abu Dhabi. Available at: https://portals.iucn.org/library/node/49130 (accessed 29 July 2025).

Cox, M., Arnold, G. and Villamayor Tomas, S. (2010) A review of design principles for community-based natural resource management. *Ecology and Society* 15(4).

Crawford, B., Balgos, M. and Pagdilao. C.R. (eds) (2000) *Community-based Marine Sanctuaries in the Philippines: A Report on Focus Group Discussions*. Coastal Management Narragansett, RI, USA and Coastal Resources Centre and Philippine Council for Aquatic and Marine Research and Development, Laguna, Philippines.

Daily, G.C. (ed.) (1997) *Nature's Services: Societal Dependence on Natural Ecosystems*. Island Press, Washington, DC, USA.

Davis, J. (2017) Vertical zoning of MPAs: when it is appropriate, when it is not, and how science is changing our understanding. Available at: https://octogroup.org/news/vertical-zoning-mpas-when-it-appropriate-when-it-not-and-how-science-changing-our/ (accessed 29 July 2025).

Daw, T.M., Hicks, C., Brown, K., Chaigneau, T., Januchowski-Hartley, F. *et al.* (2016) Elasticity in ecosystem services: exploring the variable relationship between ecosystems and human well-being. *Ecology and Society* 21(2).

Dawson, N.M., Coolsaet, B., Sterling, E.J., Loveridge, R., Gross-Camp, N.D. *et al.* (2021) The role of Indigenous peoples and local communities in effective and equitable conservation. *Ecology and Society* 26(3).

Day. J., Dudley, N., Hockings, M., Holmes, G., Laffoley, D. *et al.* (2012) *Guidelines for applying the IUCN Protected Area Management Categories to Marine Protected Areas*. IUCN, Gland, Switzerland.

Day, J., Dudley, N., Hockings, M., Holmes, G., Laffoley, D. *et al.* (eds). (2019) *Guidelines for Applying The IUCN Protected Area Management Categories to Marine Protected Areas*, 2nd edn. IUCN, Gland. Switzerland.

Devillers, R., Pressey, R.L., Grech, A., Kitinger, J.N., Edgard, G.J. *et al.* (2014) Reinventing residual reserves in the sea: are we favouring ease of establishment over need for protection? *Aquatic Conservation: Marine and Freshwater Ecosystems* 25(4): 480–504.

De Young, C., Charles, A. and Hjort, A. (2008) *Human Dimensions of the Ecosystem Approach to Fisheries: An Overview of Context, Concepts, Tools and Methods*. Fisheries Technical Paper No. 489. FAO, Rome, Italy.

Díaz, S., Demissew, S., Carabias, J., Joly, C., Lonsdale, M. *et al.* (2016) The IPBES Conceptual Framework – connecting nature and people. Current Opinion in Environmental Sustainability 14: 1–16.

Dichmont, C.M., Ellis, N., Bustamante, R.H., Deng, R., Tickell, S. *et al.* (2013) Evaluating marine spatial closures with conflicting fisheries and conservation objectives. *Journal of Applied Ecology* 50: 1060–1070.

Die, D. (2009) Fisheries management plans. In: Cochrane, K.L. and Garcia, S.M. (eds) *A Fishery Managers' Guidebook*, 2nd edn. Wiley-Blackwell Chichester, UK and FAO, Rome, Italy, pp. 425–444.

Dominguez, L. and Luoma, C. (2020) Decolonising conservation policy: how colonial land and conservation ideologies persist and perpetuate indigenous injustices at the expense of the environment. *Land* 9: 65.

Donohue, I., Hillebrand, H., Montoya, J.M., Petchey, O.L., Pimm, S.L. *et al.* (2016) Navigating the complexity of ecological stability. *Ecology Letters* 19: 1172–1185.

Dudley, N. (ed.) (2008) *Guidelines for Applying Protected Area Management Categories*. |UCN, Gland, Switzerland.

Dudley, N. and Parish, J. (2006) *Closing the Gap. Creating Ecologically Representative Protected Area Systems: A Guide to Conducting the Gap Assessments of Protected Area Systems for the Convention on Biological Diversity*. Secretariat of the Convention on Biological Diversity, Montreal, Canada.

Edgar, G.J., Stuart-Smith, R.D., Willis, T.J., Kininmonth, S., Baker, S.C. *et al.* (2014) Global conservation outcomes depend on marine protected areas with five key features. *Nature* 506: 216–220.

Ehler, C. and Douvere, F. (2009) *Marine Spatial Planning: A Step-by-step Approach Toward Ecosystem-based Management*. Intergovernmental Oceanographic Commission and Man and the Biosphere Programme, Paris, France.

European Parliament Committee on Fisheries (2012) Report on Small-Scale Coastal Fishing, Artisanal Fishing and the Reform of the Common Fisheries Policy. Available at: www.europarl.europa.eu/doceo/document/A-7-2012-0291_EN.html (accessed 29 July 2025).

Ervin, J., Mulongoy, K.J., Lawrence, K., Game, E., Sheppard, D. *et al.* (2010) *Making Protected Areas Relevant: A Guide to Integrating Protected Areas into Wider Landscapes, Seascapes and Sectoral Plans and Strategies*. CBD Technical Series No. 44. Convention on Biological Diversity, Montreal, Canada.

Fancy, S.G., Gross, J.E. and Carter, S.L. (2008) Monitoring the condition of natural resources in US national parks. *Environmental Monitoring and Assessment* 151: 161–174.

FAO (1989) *Aquaculture Systems and Practices: A Selected Review*. FAO, Rome, Italy.

FAO (1995) *Code of Conduct for Responsible Fisheries*. FAO, Rome, Italy.

FAO (1996a) *Precautionary Approach to Fisheries. 1. Guidelines on the Precautionary Approach to Capture Fisheries and Species Introductions*. FAO Fisheries Technical Paper, 350/1. FAO, Rome, Italy.

FAO (1996b) *Integration of Fisheries into Coastal Areas Management*. FAO, Rome, Italy.

FAO (1996c) *Precautionary Approach to Capture Fisheries and Species Introductions. 2.* FAO, Rome, Italy.

FAO (1997) *Fisheries Management*. FAO, Rome, Italy.

FAO (1999) *Indicators for Sustainable Development of Marine Capture Fisheries*. FAO, Rome, Italy.

FAO (2000) *Fisheries Management. 1. Conservation and Management of Sharks*. FAO, Rome, Italy.

FAO (2003) *Fisheries Management. 2. The Ecosystem Approach to Fisheries*. FAO, Rome, Italy.

FAO (2008) Report of the FAO Workshop on Vulnerable Ecosystems and Destructive Fishing in Deep-Sea Fisheries, Rome, 26–29 June. FAO, Rome, Italy.

FAO (2009a) Best practices in ecosystem modelling for informing an ecosystem approach to fisheries. In: *Fisheries Management. 2. The Ecosystem Approach to Fisheries*. FAO, Rome, Italy.

FAO (2009b) *International Guidelines for the Management of Deep-Sea Fisheries In the High Seas*. FAO, Rome, Italy.

FAO (2009c) The human dimensions of the ecosystem approach to fisheries. In: *Fisheries Management. 2. The Ecosystem Approach to Fisheries*. FAO, Rome, Italy.

FAO (2011) *Fisheries Management. 4. Marine Protected Areas and Fisheries*. FAO, Rome, Italy.

FAO (2015) *Voluntary Guidelines for Securing Sustainable Small-Scale Fisheries in the Context of Food Security and Poverty Eradication*. FAO, Rome, Italy.

FAO (2019a) Report of the Expert Meeting on Other Effective Area-Based Conservation Measures in the Marine Capture Fishery Sector, Rome, 7–10 May. FAO, Rome, Italy.

FAO (2019b) FAO Strategy on Mainstreaming Biodiversity across Agricultural Sectors. Hundred and Sixty-Third Session of the FAO Council, 2–6 December, Rome. FAO, Rome, Italy.

FAO (2021) *Biodiversity Mainstreaming Across Fisheries and Aquaculture*. FAO, Rome, Italy.

FAO (2022a) *A Handbook for Identifying, Evaluating and Reporting Other Effective Area-Based Conservation Measures in Marine Fisheries*. FAO, Rome, Italy.

FAO (2022b) Report of the Forty-Fourth Session of the General Fisheries Commission for the Mediterranean (GFCM), 2–6 November, Rome. GFCM Report No. 44. FAO, Rome, Italy.

FAO (2023) Report of the Expert Meeting on Fisheries-Related Other Effective Area-Based Conservation Measures in the Mediterranean. Available at: https://doi.org/10.4060/cc4870en (accessed 29 July 2025).

FAO/ILO/IMO (2012) *Safety Recommendations for Decked Fishing Vessels of Less than 12 Metres in Length and Undecked Fishing Vessels*. FAO, Rome, Italy.

FAO and SCBD (2023) Report of the Sustainable Ocean Initiative Capacity Building Workshop for the Wider Caribbean and Central America on Other Effective Area-Based Conservation Measures in the Marine Fishery Sector. Available at: https://doi.org/10.4060/cc8058en (accessed 29 July 2025).

FAO and UNEP (2010) Report of the FAO/UNEP Expert Meeting on Impacts of Destructive Fishing Practices, Unsustainable Fishing, and Illegal, Unreported and Unregulated (IUU) Fishing on Marine Biodiversity and Habitats. FAO, Rome, Italy..

Feydel, S. and Bonneuil, C. (2015) *Prédation. Nature, Le Nouvel Eldorado de la Finance*. Editions La Découverte, Paris, France.

Field, S.A., Tyre, A.J., Jonzen, N., Rhodes, J.R. and Possingham, H.P. (2004) Minimizing the cost of environmental management decisions by optimizing statistical thresholds. *Ecology Letters* 7: 669–675.

Field, S.A., Tyre, A.J. and Possingham, H.P. (2005) Optimizing allocation of monitoring effort under economic and observational constraints. *Journal of Wildlife Management* 69: 473–482.

Field, S.A., O'Connor, P.J., Tyre, A.J. and Possingham, H.P. (2007) Making monitoring meaningful. *Austral Ecology* 32: 485–491.

Fischer, S.H., De Oliveira, J.A., Mumford, J.D. and Kell, L.T. (2023) Risk equivalence in data-limited and data-rich fisheries management: an example based on the ICES advice framework. *Fish and Fisheries* 24(2): 231–247.

Fisher, J., Jorgensen, J., Josupeit, H., Kalikoski, D. and Lucas, C.M. (2015) *Fishers' Knowledge and the Ecosystem Approach to Fisheries. Applications, Experience and Lessons in Latin America.* FAO Fisheries and Aquaculture Technical Paper 591. FAO, Rome, Italy.

Fletcher, R. (2008) Implementing an ecosystem approach to fisheries management: lessons learned from applying a practical EAFM framework in Australia and the Pacific. In: Bianchi, G. and Skjoldal, H.R. (eds) *The Ecosystem Approach to Fisheries.* CABI, Wallingford, UK and FAO, Rome, Italy, pp. 112–124.

Fletcher, W.J. and Bianchi, G. (2014) The FAO-EAF Toolbox: making the Ecosystem Approach accessible to all fisheries. *Ocean and Coastal Management* 90: 20–26.

Fletcher, W.J., Shaw, J., Metcalf, S.J. and Gaughan, D.J. (2010) An Ecosystem Based Fisheries Management framework: the efficient, regional-level planning tool for management agencies. *Marine Policy* 34: 1226–1238.

Fletcher, W.J., Gaughan, D.J., Metcalf, S.J. and Shaw, J. (2012) Using a regional level, risk-based framework to cost effectively implement Ecosystem Based Fisheries Management (EBFM). In: Kruse, G., Browman, H., Cochrane, K. *et al.* (eds) *Global Progress on Ecosystem-Based Fisheries Management.* Alaska Sea Grant College Program, pp. 129–146. Available at: https://doi.org/10.4027/gpebfm. (2012)07 (accessed 30 July 2025).

Foster, N.L., Rees, S., Langmead, O., Griffiths, C., Oates, J. and Attrill, M.J. (2017) Assessing the ecological coherence of a marine protected area network in the Celtic Seas. *Ecosphere* 8(2): e01688.

Fox, H.E., Holtzman, J.L., Haisfield, K.M., McNally, C.G., Cid, G.A. *et al.* (2014) How are our MPAs doing? Challenges in assessing global patterns in marine protected area performance. *Coastal Management* 42: 207–226.

Freestone, D., Cicin Sain, B., Hewawasam, E. and Hamon, G. (2010) *Draft Policy Brief on: Achieving Integrated Ecosystem-Based Ocean and Coastal Management.* Global Forum on Oceans, Coasts and Islands. Global Oceans Conference, 3–7 May, Paris.

Friedman, K., Garcia, S.M. and Rice, J. (2018) Mainstreaming biodiversity in fisheries. *Marine Policy* 95: 209–220.

Fulton, B., Punt, A., Dichmont, C., Gorton, B., Sporcic, M. *et al.* (2016) Developing risk equivalent data-rich and data-limited harvest strategies. *Fisheries Research* 183: 574–587.

Fulton, E.A., Smith, A.D.M. and Punt, A.E. (2005) Which ecological indicators can robustly detect effects on fishing? *ICES Journal of Marine Science* 62: 540–551.

Fulton, E.A., Bax, N.J., Bustamante, R.H., Dambacher, J.M., Dichmont, C. *et al.* (2015) Modelling marine protected areas: insights and hurdles. *Philosophical Transactions of the Royal Society B* 370: 20140278.

Gallacher, J., Simmonds, N., Fellowes, H., Brown, N., Gill, N. *et al.* (2016) Evaluating the success of a marine protected area: a systematic review approach. *Journal of Environmental Management* 183: 280–293.

Garcia, S.M. (2009) Governance, science and society. In: Grafton, Q.R., Hilborn, R., Squires, D., Tait, M. and Williams, M. (eds) *Handbook of Marine Fisheries Conservation and Management.* Oxford University Press, Oxford, pp. 87–98.

Garcia, S.M. and Charles, A.T. (2007) Fishery systems and linkages: from clockwork to soft watches. *ICES Journal of Marine Science* 64(4): 580–587.

Garcia, S.M. and Cochrane, K. (2005) Ecosystem approach to fisheries: a review of implementation guidelines. *ICES Journal of Marine Science* 62(3): 311–318.

Garcia, S.M. and Rice J. (eds) (2024) *Area-Based Management Tools and Fisheries, A Comprehensive Review.* Cambridge Scholars Publishing, Newcastle upon Tyne, UK.

Garcia, S.M., Zerbi, A., Alliaume, C., DoChi, T. and Lasserre, G. (2003) *The Ecosystem Approach o Fisheries. Issues, Terminology, Principles, Institutional Foundations, Implementation and Outlook.* FAO Fisheries Technical Paper 443. FAO, Rome, Italy.

Garcia, S.M., Allison, E.H., Andrew, N., Bené, C., Bianchi, G. *et al.* (2008) *Towards Integrated Assessment and Advice in Small-Scale Fisheries. Principles and Processes.* FAO Fisheries Technical Paper 515. FAO, Rome, Italy.

Garcia, S.M., Rey-Vallette, H. and Bodiguel, C. (2009) Which indicators for what management? The challenge of connecting offer and demand of indicators. In: Cochrane, K. and Garcia, S.M. (eds) *A Fishery Managers' Handbook.* FAO, Rome, Italy and Wiley-Blackwell, Chichester, UK, pp. 303–332.

Garcia, S.M., Kolding, J., Rice, J., Rochet, M-J., Zhou, S. *et al.* (2012) Reconsidering the consequences of selective fisheries. *Science Policy Forum* 335: 1045–1047.

Garcia, S.M., Boncoeur, J. and Gascuel, D. (2013) *Les Aires Marines Protegees et La Peche: Bioecologie, Socioeconomie et Gouvernance.* Presses Universitaires de Perpignan, Perpignan, France.

Garcia, S.M., Rice, J., Friedman, K. and Himes-Cornell, A. (2019) *Identification, Assessment and Governance of Other Effective Area-Based Conservation Measures in the Marine Fishery Sector: A Background Document.* Prepared for the FAO/SCBD/IUCN-CEM-FEG Expert Meeting on OECMs in the Marine Fishery Sector, Rome, Italy, 7–10 May.

Garcia, S.M., Rice, J., Charles, A. and Diz, D. (2020a) OECMs in Marine Capture Fisheries: Brief for Policy-Makers and Managers. Fisheries Expert Group of the IUCN Commission on Ecosystem Management, Gland, Switzerland and European Bureau of Conservation and Development, Brussels, Belgium.

Garcia, S.M., Rice, J., Charles, A. and Diz, D. (2020b) OECMs in Marine Capture Fisheries: Systematic Approach to Identification, Use and Performance Assessment in Marine Capture Fisheries. Fisheries Expert Group of the IUCN Commission on Ecosystem Management, Gland, Switzerland and European Bureau of Conservation and Development, Brussels, Belgium.

Garcia, S.M., Rice, J., Charles, A. and Diz, D. (2021) *OECMs in Marine Capture Fisheries: Systematic Approach to Identification, Use and Performance Assessment in Marine Capture Fisheries (Version 2).* Fisheries Expert Group of the IUCN Commission on Ecosystem Management, Gland, Switzerland and European Bureau of Conservation and Development, Brussels, Belgium.

Garcia, S.M., Rice, J., Himes-Cornell, A., Friedman, K.J., Charles, A. *et al.* (2022) OECMs in marine capture fisheries: key implementation issues of governance, management, and biodiversity. *Frontiers in Marine Science* 9: 920051.

Garcia, S.M., Rice, J., Link, J., Sowman, M., Charles, A. *et al.* (2023) *Area-based Management Tools and Fisheries: History, Definitions, Roles, Typologies, Tensions, Synergies, Trade-Offs, and Effectiveness.* Available at: https://ebcd.org/wp-content/uploads/2023/04/2023-04-08-FEG-ABMT-Master-15-1.pdf (accessed 30 July 2025).

GFCM (2024) *Emerging Novelties Regarding the OECM Reporting Process.* Report on 2023–2024 advice from subsidiary bodies. FAO, Rome, Italy.

Gill, D.A., Mascia, M.B., Ahmadia, G.N., Glew, L., Lester, S.E. *et al.* (2017) Capacity shortfalls hinder the performance of marine protected areas globally. *Nature* 543(7647): 665–669.

Gillet, R. (2017) *A Review of Special Management Areas in Tonga.* FAO Fisheries and Aquaculture Circular 1137. Apia, Samoa.

Gilman, E., Kaiser, M.J. and Chaloupka, M. (2019) Do static and dynamic marine protected areas that restrict pelagic fishing achieve ecological objectives? *Ecosphere* 10(12): e02968.

Goulletquer, P., Gros, P., Bœuf, G. and Weber, J. (2013) *Biodiversité en Environnement Marin.* Editions Quae, Versailles, France.

Govan, H., Aalbersberg, W., Tawake, A. and Parks, J. (2008) *Locally-Managed Marine Areas: A Guide for Practitioners.* Locally-Managed Marine Area Network. Available at: www.LMMAnetwork.org (accessed 30 July 2025).

Govan, H., Lalavanua, W. and Steenbergen, D.J. (2024) Coastal fisheries governance in the Pacific Islands: the evolution of policy and the progress of management-at-scale. In: Nakamura, J., Chuenpagdee, R. and Jentoft, S. (eds) *Implementation of the Small-Scale Fisheries Guidelines.* MARE Publication Series, vol 28. Springer, Cham, Switzerland.

Government of Canada (2022) *Guidance for Recognizing Marine Other Effective Area-Based Conservation Measures.* Document K1A 0E6. Fisheries and Oceans Canada, Ottawa, Ontario. Available at: www.dfo-mpo.gc.ca/oceans/publications/oecm-amcepz/guidance-directives-2022-eng.html (accessed 30 July 2025).

Grafton, Q.R., Hilborn, R., Squires, D., Tait, M. and Williams, M. (eds) (2009) *Handbook of Marine Fisheries Conservation and Management.* Oxford University Press, Oxford, UK.

Graham, J., Amos, B. and Plumptre, T. (2003) *Principles for Good Governance in the 21st Century.* Policy Brief 15. Institute of Governance, Ottawa, Canada.

Green, S.J., White, A.T., Christie, P., Kilarski, S., Meneses, A.B.T. *et al.* (2011) Emerging marine protected area networks in the coral triangle: lessons and way forward. *Conservation and Society* 9(3): 173–188.

Griffiths, R.C., Robles, R., Coppola, S.R, and Camiñas, J.A. (2007) *Is There a Future for Artisanal Fisheries in the Western Mediterranean?* COPEMED Document. FAO, Rome, Italy.

Grober-Dunsmore, R., Wooninck, L., Field, J., Ainsworth, C., Beets, J. *et al.* (2009) Vertical zoning in marine protected areas ecological considerations for balancing pelagic fishing with conservation of benthic communities features. *Fisheries* 33(12): 598–610.

Grorud-Colvert, K., Claudet, J., Tissot, B.N., Caselle, J.E., Carr, M.H. *et al.* (2014) Marine protected area networks: assessing whether the whole is greater than the sum of its parts. *PLoS ONE* 9: 1–7.

Halpern, B., Regan, H., Possingham, H. and McCarthy, M. (2005) Rejoinder: uncertainty and decision-making. *Ecology Letters* 9: 13–14.

Hattam, C., Atkins, J.P., Beaumont, N., Borgen, T., Bohnke-Henrich, A. *et al.* (2015) Marine ecosystem services: linking indicators to their classification. *Ecological Indicators* 49: 61–75.

HELCOM (2022) *Report from the HELCOM Regional Workshop on OECMs.* Baltic Marine Environment Protection Commission Helsinki Commission. Online meeting, 3–4 March.

Hilborn, R. and Sinclair A.R.E. (2021) Biodiversity protection in the 21st century needs intact habitat and protection from overexploitation whether inside or outside parks. *Conservation Letters* 14: e12830.

Hilborn, R. and Walters, C.J. (1992) *Quantitative Fisheries Stock Assessment: Choice, Dynamics And Uncertainty.* Chapman and Hall, London.

Hilborn, R., Stokes, K., Maguire, J.-J., Smith, T., Botsford, L.W. *et al.* (2004) When can marine reserves improve fisheries management? *Ocean and Coastal Management* 47: 197–205.

Hilborn, R., Agostini, V., Chaloupka, M., Garcia, S.M., Gerber, L.R. *et al.* (2021a) Area-based management of blue water fisheries: current knowledge and research needs. *Fish and Fisheries* 23, 492–518.

Hilborn, R., Akserlrud, C.A., Peterson, H. and Whitehouse, G.A. (2021b) Trade-off between biodiversity and sustainable fish harvest with area-based management. *ICES Journal of Marine Science* 78(6): 2271–2279.

Hiltz E., Fuller, M.D. and Mitchell, J. (2018) *Disco Fan Conservation Area: A Canadian Case Study. PARKS* 24 (Special Issue): 17–29.

Himes, A.H. (2007) Performance indicators in MPA management: using questionnaires to analyze stakeholder preferences. *Ocean and Coastal Management* 50(5-6): 329–351.

Himes-Cornell, A., Lechuga Sanchez, J.F., Potter, C., McKean, C., Rice, J. *et al.* (2022) Reaching global marine biodiversity conservation goals with area-based fisheries management: a typology-based evaluation. *Frontiers in Marine Science* 9: 932283.

Hindson, J., Hoggarth, D., Krishna, M., Mees, C.C. and O'Neill, C. (2005) *How to Manage a Fishery. A Simple Guide to Writing a Fishery Management Plan.* Marine Resources Assessment Group, London, UK.

Hobday, A.J., Smith, A.D.M., Stobutzki, I.C., Bulman, C., Daley, R. *et al.* (2011) Ecological risk assessment for the effects of fishing. *Fisheries Research* 108(2-3): 372–384.

Hockings, M. (1998) Evaluating management of protected areas: integrating planning and evaluation. *Environmental Management* 22(3): 337–345.

Holland, D.S. (2010) *Management Strategy Evaluation and Management Procedures: Tools for Rebuilding and Sustaining Fisheries.* OECD Food, Agriculture and Fisheries Working Paper 25. OECD Publishing. Available at: https://doi.org/10.1787/5kmd77jhvkjf-en (accessed 30 July 2025).

Huntley, B.J. and Redford, K.H. (2014) *Mainstreaming Biodiversity in Practice: A STAP Advisory Document.* Global Environment Facility, Washington, DC, USA.

ICES (2002) *Report of the Study Group on the Further Development of the Precautionary Approach to Fisheries Management.* Lisbon, 4–8 March.

ICES (2021) ICES/IUCN-CEM FEG Workshop on Testing OECM Practices and Strategies (WKTOPS). Available at: https://doi.org/10.17895/ices.pub.8135 (accessed 30 July 2025).

ICES (2023a) *Workshop to Evaluate Long-Term Biodiversity/Ecosystem Benefits of NEAFC Closed and Restricted Areas (WKECOVME).* Available at: https://doi.org/10.17895/ices.pub.24198771.v1 (accessed 30 July 2025).

ICES (2023b) *NEAFC request on Other Effective Area-Based Conservation Measures in relation to long-term biodiversity/ecosystem benefits of NEAFC's closed areas and areas restricted to bottom fishing.* Available at: https://ices-library.figshare.com/articles/report/NEAFC_request_on_Other_Effective_Area-Based_Conservation_Measures_in_relation_to_long-term_biodiversity_ecosystem_benefits_of_NEAFC_s_closed_areas_and_areas_restricted_to_bottom_fishing/24230083?file=43679682 (accessed 30 July 2025).

Inoni, O.E. and Oyaide, W.J. (2007) Socio-economic analysis of artisanal fishing in the south agro-ecological zone of Delta State, Nigeria. *Agricultura Tropica et Subtropica* 40(4).

INTOSAI WGEA (2007a) *Auditing Biodiversity: Guidance for Supreme Audit Institutions.* Available at: https://environmental-auditing.org/media/3235/eng07pr_fs_guidebiodiversity.pdf (accessed 30 July 2025).

INTOSAI WGEA (2007b) *Evolution and Trends in Environmental Auditing.* Available at: www.environmental-auditing.org/media/wflpu10u/evolution-and-trends-in-environmental-auditing.pdf (accessed 30 July 2025).

INTOSAI WGEA (2019) *Environmental Audit and the Sustainable Development Goals: A Discussion Paper.* Available at: https://www.environmental-auditing.org/media/113691/21h-wgea_sdgs_18-sep-2019.pdf (accessed 30 July 2025).

IPBES (2016) *The Methodological Assessment Report on Scenarios and Models of Biodiversity and Ecosystem Services.* Intergovernmental Science-Policy Platform on Biodiversity and Ecosystem Services. Available at: https://ipbes.net/assessment-reports/scenarios (accessed 30 July 2025).

IPBES (2018a) *The IPBES Regional Assessment Report on Biodiversity and Ecosystem Services for the Americas.* Secretariat of the Intergovernmental Science-Policy Platform on Biodiversity and Ecosystem Services, Bonn, Germany.

IPBES (2018b) *The IPBES Regional Assessment Report on Biodiversity and Ecosystem Services for Asia and the Pacific.* Secretariat of the Intergovernmental Science-Policy Platform on Biodiversity and Ecosystem Services, Bonn, Germany.

IPBES (2018c) *The IPBES Regional Assessment Report on Biodiversity and Ecosystem Services for Africa.* Secretariat of the Intergovernmental Science-Policy Platform on Biodiversity and Ecosystem Services, Bonn, Germany.

IPBES (2019) *Global Assessment Report on Biodiversity and Ecosystem Services of the Intergovernmental Science-Policy Platform on Biodiversity and Ecosystem Services.* Secretariat of the Intergovernmental Science-Policy Platform on Biodiversity and Ecosystem Services, Bonn, Germany.

IPBES (2022a) *Thematic Assessment Report on the Sustainable Use of Wild Species of the Intergovernmental Science-Policy Platform on Biodiversity and Ecosystem Services.* Secretariat of the Intergovernmental Science-Policy Platform on Biodiversity and Ecosystem Services, Bonn, Germany.

IPBES (2022b) *Summary for Policy-makers of the Methodological Assessment Report on the Diverse Values and Validation of Nature of the International Science-Policy Platform om Biodiversity and Ecosystem Services.* Secretariat of the Intergovernmental Science-Policy Platform on Biodiversity and Ecosystem Services, Bonn, Germany.

IPBES (2024) *Thematic Assessment Report on the Interlinkages among Biodiversity, Water, Food and Health of the Intergovernmental Science-Policy Platform on Biodiversity and Ecosystem Services.* Secretariat of the Intergovernmental Science-Policy Platform on Biodiversity and Ecosystem Services, Bonn, Germany.

IUCN-WCPA (2019) Recognising snd Reporting Other Effective Area-Based Conservation Measures. IUCN, Gland, Switzerland.

IUCN-WCPA (2020) Potential Contribution of "Other-effective area-based conservation measures" to Achieving Aichi Target 11 in Southern and Eastern Mediterranean Countries. Report on the Regional Workshop, Tunis, Tunisia, 10–11 February. IUCN Gland, Switzerland.

Ivanic, K-Z., Stolton, S., Figueroa Arango, C. and Dudley, N. (2020) *Protected Areas Benefits Assessment Tool + (PA-BAT+): A Tool to Assess Local Stakeholder Perceptions of the Flow of Benefits from Protected Areas.* IUCN, Gland, Switzerland.

Jacobsen, N.S., Burgess, M.G. and Andersen, K.H. (2016) Efficiency of fisheries is increasing at the ecosystem level. *Fish and Fisheries* 18, 199–211.

Jennings, S. and Kaiser, M.J. (1998) The effects of fishing on marine ecosystems. *Advances in Marine Biology* 34, 201–212.

Jentoft, S. (2017) Small-scale fisheries within maritime spatial planning: knowledge integration and power. *Journal of Environmental Policy and Planning* 19: 266–278.

Johannes, R.E. (1978) Traditional marine conservation methods in Oceania and their demise. *Annual Review in Ecology and Systematics* 9: 349–364.

Johannes, R.E. (1982) Traditional conservation methods and protected marine areas in Oceania. *Ambio* 11(5): 258–261.

Johnson, D.E., Barrio Froján, C., Diz, D., Ferreira, M.A. and Halpin, P.N. (2024) Other effective conservation measures in the marine environment: the policy-makers' silver bullet for meeting global conservation ambitions? In: Phillips, M.R., Al-Naemi, S. and Duarte, C.M. (eds) *Coastlines under Global Change: Proceedings from the International Coastal Symposium (ICS) 2024 (Doha, Qatar). Journal of Coastal Research* 113(Special Issue): 16–20.

Jonas, H.D. and Jonas H.C. (2019) Are "conserved areas" conservation's most compelling story? *PARKS* 25(2): 103–108.

Jonas, H.D., Ahmadia, G.N., Bingham. H.C., Briggs, J., Butchart, S.H.M. *et al.* (2021) Equitable and effective area-based conservation: towards the conserved areas paradigm. *PARKS* 27(1): 71–84.

Jonas, H.D., Wood, P. and Woodley, S. (eds) (2024) *Guidance on Other Effective Area-Based Conservation Measures (OECMs)*. Good Practice Series 36. IUCN, Gland, Switzerland.

Jorgensen, L.L., Bakke, G. and Hoel, A.H. (2020) Responding to global warming: new fisheries management measures in the Arctic. *Progress in Oceanography* 188: 102423.

Kaiser, M.J., Collie, J.S., Hall, S.J., Jennings, S. and Poiner, I.R. (2003) Impacts of fishing gear on marine benthic habitats: In: Sinclair, M. and Valdimarsson, G. (eds) *Responsible Fisheries in the Marine Ecosystem*. FAO, Rome, Italy and CABI, Wallingford, UK, pp. 197–217.

Kaplan, D.M. (2009) Fish life histories and marine protected areas: an odd couple? *Marine Ecology Progress* 377: 213–225.

Kearney, R. and Farebrother, G. (2012) Marine management: expand Australia's sustainable fisheries. *Nature* 482: 162.

Keith, D.A., Martin, T.G., McDonald-Madden, E. and Walters, C. (2011) Uncertainty and adaptive management for biodiversity conservation. *Biological Conservation* 144(4): 1175–1178.

Kemp, J., Jenkins, G.P., Smith, D.C. and Fulton, E.A. (2012) Measuring the performance of spatial management in marine protected areas. *Oceanography and Marine Biology: An Annual Review* 50: 287–314.

Kenchington, R., Kaiser, M.J. and Boerder, K. (2018) Marine Protected Areas, fishery closures and fisheries rebuilding. In: Garcia, S.M., Ye, Y., Rice, J. and Charles, T. (eds) *Rebuilding of Marine Fisheries. Part 2: Case Studies*. FAO Fisheries and Aquaculture Technical Paper 630(2). FAO, Rome, Italy, pp. 182–206.

Klara Loch, T. and Riechers, M. (2021) Integrating indigenous and local knowledge in management and research on coastal ecosystems in the Global South: a literature review. *Ocean and Coastal Management* 212: 105821.

Knoops, P. (1995) *Background Material and Guidelines for the Chartering of Industrial Fishing Vessels in Cape Verde. Improvement of the Legal Framework for Fisheries Cooperation, Management, and Development In Coastal States of West Africa*. Project GRP/RAF/302/EEC Document 22. FAO, Dakar, Africa.

Kooiman, J. (2005) Part IV. Principles for fisheries governance. In: Koiman, J., Bavink, M., Jentoft, S. and Pullin, R. (eds) *Fish for Life. Interactive Governance for Fisheries*. Amsterdam University Press, Amsterdam, The Netherlands, pp. 241–302.

Kothari, A., Corrigan, C., Jonas, H., Neumann, A. and Shrumm, H. (eds) (2012) *Recognising and Supporting Territories and Areas Conserved By Indigenous Peoples And Local Communities: Global Overview and National Case Studies*. Technical Series 64. SCBD, ICCA Consortium, Kalpavriksh, and Natural Justice. Montreal, Canada.

Kremen, C. (2005) Managing ecosystem services: what do we need to know about their ecology? *Ecology Letters* 8: 468–479.

Langlois, J., Fréon, P., Delgenes, J.-P., Steyer, J.-P. and Hélias, A. (2014) New methods for impact assessment of biotic-resource depletion in life cycle assessment of fisheries: theory and application. *Journal of Cleaner Production* 73: 63–71.

Larsen, A., Meng, K. and Kendall, B. (2019) Causal analysis in control-impact ecological studies with observational data. *Methods in Ecology and Evolution* 10: 924–934.

Leite, L., Thiao, D., Westlund, L. and Zahri, Y. (2019) *Participatory Monitoring and Evaluation in Marine Protected Areas: Experiences from North and West Africa*. FAO Fisheries and Aquaculture Circular 1173. FAO, Rome, Italy.

Letten, A.D., Ke, P.-J. and Fukami, T. (2017) Linking modern coexistence theory and contemporary niche theory. *Ecological Monographs* 87(2): 161–177.

Lin, C.Y., Wang, S.P., Chiang, W.C., Griffiths, S. and Yeh, H.M. (2020) Ecological risk assessment of species impacted by fisheries in waters off eastern Taiwan. *Fisheries Management and Ecology* 27(4): 345–356.

Lindenmayer, D.B. and Likens, G.E. (2010) The science and application of ecological monitoring. *Biological Conservation* 143: 1317–1328.

Link, J. (2018) System-level optimal yield: increased value, less risk, improved stability, and better fisheries. *Canadian Journal of Fisheries and Aquatic Sciences* 75: 1–16.

Link, J., Watson, R.A., Pranovic, F. and Libralato, S. (2020) Comparative production of fisheries yields and ecosystem overfishing in African Large Marine Ecosystems. *Environmental Development* 36: 100529.

Link, J.S. (2010) *Ecosystem-based Fisheries Management. Confronting Trade-offs.* Cambridge University Press, Cambridge, UK.

Lopes de Sousa, W., Maia Zacardi, D. and Almeida Vieira, T. (2022) Traditional ecological knowledge of fishermen: people contributing towards environmental preservation. *Sustainability* 14(9): 4899.

Lopoukhine, N. and Ferreira de Souza Dias, B. (2012) What does target 11 really mean? *PARKS* 18(1): 5–8.

Ludwig, D. and Hilborn, R. (1983) Adaptive probing strategies for age structured fish stocks. *Canadian Journal of Fisheries and Aquatic Sciences* 40(5): 559–569.

Mace, G. (2014) Whose conservation? *Science* 345: 1558–1559.

Mace, G.M., Norris, K. and Fitter, A.H. (2012) Biodiversity and ecosystem services: a multilayered relationship. *Trends in Ecology and Evolution* 27: 19–26.

Mardle, S., Pascoe, S. and Herrero, I. (2004) Management objective importance in fisheries: an evaluation using the analytic hierarchy process (AHP). *Environmental Management* 33: 1–11.

Marnewick, D., Stevens, C. and Jonas H. (eds) (2019) *A Step-by-step Methodology for Identifying, Reporting, Recognising and Supporting Other Effective Area-based Conservation Measures.* IUCN, Gland, Switzerland.

Marnewick, D., Stevens, C., Antrobus-Wuth, R., Theron, N., Wilson, N. *et al.* (2020) *Assessing the Extent of OECMs in South Africa: Final Project Report.* BirdLife South Africa, Johannesburg, South Africa.

Mascia, M., Fox, H., Glew, L., Ahmadia, G., Agrawal, A. *et al.* (2017) A novel framework for analysing conservation impacts: evaluation, theory, and marine protected areas. *Annals of the New York Academy of Sciences* 1399: 93–115.

Maxwell, S.M., Gjerde, K.M., Conners, M.G. and Crowder, L.B. (2020) Protecting mobile marine species and habitats under climate change will require innovative and dynamic tools. *Science* 367(6475). 252–254.

McCarthy, A.H., Steadman, D., Richardson, H., Murphy, J., Benbow, S. *et al.* (2024) Destructive fishing: an expert-driven definition and exploration of this quasi-concept. *Conservation Letters* 17: e13015.

McElwee, P., Fernández-Llamazares, Á., Aumeeruddy-Thomas, Y., Babai, D., Bates, P. *et al.* (2020) Working with Indigenous and local knowledge (ILK) in large-scale ecological assessments: reviewing the experience of the IPBES Global Assessment. *Journal of Applied Ecology* 57(9): 1666–1676.

McMillan, L.J. and Prosper, K. (2016) Remobilizing netukulimk: indigenous cultural and spiritual connections with resource stewardship and fisheries management in Atlantic Canada. *Reviews in Fish Biology and Fisheries* 26(4): 629–647.

MEA (2005) *Ecosystems and Human Wellbeing. Our Human Planet: Summary for Decision-makers.* Millennium Ecosystem Assessment. Island Press, Washington, DC, USA.

Meehan, M.C., Ban, N.C., Devillers, R., Singh, G.G. and Claudet. J. (2020) How far have we come? A review of MPA network performance indicators in reaching qualitative elements of Aichi Target 11. *Conservation Letters* 13: e12746.

Mesnildrey, L., Gascuel, D. and Le Pape, O. (2013) Integrating marine protected areas in fisheries management systems: some criteria for ecological efficiency. *Aquatic Living Resources* 26(2): 159–170.

Milner-Gulland, E.J., Garcia, S., Arlidge, W., Bull, J., Charles, A. *et al.* (2018) Translating the terrestrial mitigation hierarchy to marine megafauna by-catch. *Fish and Fisheries* 19: 547–561.

Misund, O.A., Kolding, J. and Fréon, P. (2002) Fish capture devices in industrial and artisanal fisheries and their influence on management. In: Hart, P.J. . and Reynolds, J.D. (eds) *Handbook of Fish Biology and Fisheries*, vol. II. Blackwell Science, London, pp. 13–36.

Moldan, B., Billharz, S. and Matravers, R. (eds) (1997) *Sustainability Indicators: A Report on the Project on Indicators of Sustainable Development.* SCOPE Report 58. John Wiley and Sons, Chichester, UK.

Mrozowski, S.A. (1999) Colonization and the commodification of nature. *International Journal of Historical Archaeology* 3: 153–166.

Murillo, F.J., Muñoz, P.D., Cristobo, J., Rios, P., González, C. *et al.* (2012) Deep-sea sponge grounds of the Flemish Cap, Flemish Pass and the Grand Banks of Newfoundland (northwest Atlantic Ocean): distribution and species composition. *Marine Biology Research* 8: 842–854.

Murillo, F.J., Kenchington, E., Koen-Alonso, M., Guijarro, J., Kenchington, T.J., Sacau, M., *et al.* (2020a) Mapping benthic ecological diversity and interactions with bottom-contact fishing on the Flemish Cap (northwest Atlantic). *Ecological Indicators* 112: 106135.

Murillo, F.J., Weigel, B., Kenchington, E. and Bouchard Marmen, M. (2020b) Marine epibenthic functional diversity on Flemish Cap (northwest Atlantic) – identifying trait responses to the environment and mapping ecosystem functions. *Diversity and Distributions* 26(4): 460–478.

NACA/FAO (2000) *Report of the Conference on Aquaculture in the Third Millennium*. Conference on Aquaculture in the Third Millennium, 20–25 February Bangkok, Thailand. NACA, Bangkok, Thailand and FAO, Rome, Italy.

NAFO (2021) *Report of the NAFO Joint Commission-Scientific Council Working Group on the Ecosystem Approach Framework to Fisheries Management (WG-EAFFM) Meeting*. NAFO COM-SC Document 21-03. NAFO, Halifax, Canada. Available at: www.nafo.int/Portals/0/PDFs/COM-SC/2021/com-scdoc21-03.pdf (accessed 30 July 2025).

NAFO (2023) *Northwest Atlantic Fisheries Organization Scientific Council Reports. PART C: Report of the Scientific Council Meeting, 2–15 June*. NAFO, Halifax, Canada. Available at: www.nafo.int/Portals/0/PDFs/rb/2023/redbook2023_final.pdf (accessed 30 July 2025).

NAFO/COM-SC (2023) *Report of the NAFO Joint Commission-Scientific Council Working Group on Ecosystem Approach Framework to Fisheries Management Meeting, 20–22 July*. Document 23-04. NAFO, Halifax, Canada. Available at: www.nafo.int/Portals/0/PDFs/COM-SC/2023/com-scdoc23-04.pdf (accessed 30 July 2025).

NEAFC (2023a) *Report of the 42nd Annual Meeting of the North-East Atlantic Fisheries Commission, London, 14–17 November*. Available at: www.neafc.org/system/files/Report%20AM_2023%20Final.pdf (accessed 30 July 2025).

NEAFC (2023b) Press release from the 2023 Annual Meeting of the North-East Atlantic Fisheries Commission, London, 14–17 November.

NEAFC (2023c) *Proposal by DFG, Iceland, Norway. AM 2023-100*. Available at: www.neafc.org/system/files/AM_2023-100_Proposal-by_Nor-Isl-DFG_OECMs-NEAFC-AM-(2023)pdf (accessed 30 July 2025).

NOAA (2017) *Preparing Management Plans for Trawl Fisheries in Southeast Asia*. Report on Workshop 3 (NA15NMF4630204) and supplementary workshop (NA16NMF0080374) on multispecies maximum sustainable yield. 24–27 April, Bangkok, Thailand.

Norse, E.A., Crowder, L.B., Gjerde, K.M. Hyrenbach, D., Roberts, C. *et al.* (2005) Place-based ecosystem management in the open ocean. In: Norse, E.A. and Crowder, L.B. (eds) *Marine Conservation Biology: The Science of Maintaining the Sea's Biodiversity*. Island Press, Washington, DC, USA, pp, 302–317.

OECD (2024) *How to Adapt Fisheries to Climate Change*. A workshop of the OECD Committee on Fisheries on Climate Change and Fisheries. OECD, Paris, France. Available at: https://issuu.com/oecd.publishing/docs/how_to_adapt_fisheries_to_climate_change (accessed 30 July 2025).

O'Keefe, C.E., Cadrin S.X. and Stokesbury, K.D.E. (2014) Evaluating effectiveness of time/area closures, quotas/caps, and fleet communications to reduce fisheries bycatch. *ICES Journal of Marine Science* 71: 1286–1297.

Ounanian, K., Carballo-Cárdenas, E., van Tatenhove J.P.M., Delaney, A., Papadopoulou, K.N. and Smith, C.J. (2018) Governing marine ecosystem restoration: the role of discourses and uncertainties. *Marine Policy* 96: 136–144.

Pascual, U., Balvanera, P., Diaz, S., Pataki, G., Roth, E. *et al.* (2017) Valuing nature's contributions to people: the IPBES approach. *Current Opinion in Environmental Sustainability* 26–27: 7–16.

Perez-Alvaro, E. and Boswell, R. (2025) Integral oceans heritage of indigenous communities: its value for good health and well-being. *Social Sciences and Humanities Open* 11: 101245.

Petza, D., Maina, I., Koukourouvli, N., Dimarchopoulou, D. Akrivos, D. *et al.* (2017) Where not to fish – reviewing and mapping fisheries restricted areas in the Greek Aegean Sea. *Mediterranean Marine Science* 18(2): 310–323.

Petza, D., Chalkias, C., Koukourouvli, N., Coll, M., Vassilopoulou, V. *et al.* (2019) An operational framework to assess the value of fisheries restricted areas for marine conservation. *Marine Policy* 102: 28–39.

Petza, D., Anastopoulos, A., Kalogirou, S., Coll, M., Garcia, S.M. *et al.* (2023) Contribution of area-based fisheries management measures to fisheries sustainability and marine conservation: a global scoping review. *Reviews in Fish Biology and Fisheries* 33: 1049–1073.

Plagányi, E.E. (2007) *Models for an Ecosystem Approach to Fisheries*. FAO Fisheries Technical Paper 477. FAO, Rome, Italy.

Pollnac, R.B. (1994) Research directed at developing local organizations for people's participation in fisheries management. In: Pomeroy, R.S. (ed.) *Community Management and Common Property of Coastal Fisheries in Asia and the Pacific: Concepts, Methods and Experiences*. ICLARM, Manila, The Philippines, pp. 94–106.

Pomeroy, R.S., Parks, J.E. and Watson, L.M. (2004) *How is Your MPA Doing? A Guidebook of Natural and Social Indicators for Evaluating Marine Protected Area Management Effectiveness*. IUCN, Gland, Switzerland and WWF/NOAA, Cambridge, UK.

Pomeroy, R.S., Watson, L.M., Parks, J.E. and Cid, G.A. (2005) How is your MPA doing? A methodology for evaluating the management effectiveness of marine protected areas. *Ocean and Coastal Management* 48: 485–502.

Punt, A. and Ralston, S. (2007) A management strategy evaluation of rebuilding revision rules for overfished rockfish stocks. In: Heifetz, J., DiCosimo, J., Gharrett, A.J., Love, M.S., O'Connell, V.M. and Stanley R.D. (eds) *Biology, Assessment, and Management of North Pacific Rockfishes*. Alaska Sea Grant, University of Alaska Fairbanks, Fairbanks, USA, pp. 329–351.

Quimby, B. and Levine, A. (2018) Participation, power, and equity: examining three key social dimensions of fisheries co-management. *Sustainability* 10(9): 3324.

Rice, J., Daan, N., Gislason, H. and Pope, J. (2013) Does functional redundancy stabilize fish communities? *ICES Journal of Marine Science* 70: 734–742.

Rice, J., Garcia, S.M. and Kaiser, M. (2018) *Other Effective Area-based Conservation Measures (OEABCMs) Used in Marine Fisheries: A Working Paper*. Available at: www.cbd.int/doc/c/0689/522e/7f94ced371f a41aeee6747e5/mcb-em-2018-01-inf-04-en.pdf (accessed 30 July 2025).

Rice, J., Friedman, K., Garcia, S.M., Govan, H. and Himes-Cornell, A. (2022) A contrast of criteria for special places important for biodiversity outcomes. *Frontiers in Marine Science* 9 : 91203.

Rice J., Garcia S.M., Charles, A. and Kirkegaard, E. (2023) *Fisheries and the Targets of the Global Biodiversity Framework – Opportunities, Challenges, and Concerns*. IUCN-CEM-FEG and EBCD, Brussels, Belgium.

Rice, J., Duplisea, D.E., Hunter, K.J. and Roux, M.-J. (2025) Risk equivalent safe operating space as an inclusive framework for living resource management in a multisectoral, multicultural world. *Facets* 10: 1–18.

Rice, J.C. (2009) A generalisation of the three-stage model for advice using a Precautionary Approach in fisheries, to apply to a broadly to ecosystem properties and pressures. *ICES Journal of Marine Science* 67: 433–444.

Rice, J.C. and Houston, K.A. (2011) *Representativity and networks of Marine Protected Areas. Aquatic Conservation: Marine and Freshwater Ecosystems* 21: 649–657.

Rice, J.C. and Rochet, M.-J. (2005) A framework for selecting a suite of indicators for fisheries management. *ICES Journal of Marine Science* 62(3): 516–527.

Roberts, C.M., O'Leary, B.C., McCauley, D.J., Cury, P.M., Duarte, C.M., et al, (2017) Marine reserves can mitigate and promote adaptation to climate change. *PNAS* 114(24): 6167–6175.

Rock, J., Sima, E. and Knapen, M. (2020) What is the ocean: a sea-change in our perceptions and values? *Aquatic Conservation: Marine and Freshwater Ecosystems* 30(3): 532–539.

Rocliffe, S. (2015) Scaling and sustaining locally managed marine areas. PhD thesis, New York University, New York, USA.

Rousseau, Y., Watson, R.A., Blanchard, J. and Fulton, E.A. (2019) Evolution of global marine fishing fleets and the response of fished resources. *PNAS* 116(25): 612238–12243.

Rudnick, D., Ryan, S.J., Beier, P., Cushman, S.A., Dieffenbach, F. et al. (2012) *The Role of Landscape Connectivity in Planning and Implementing Conservation and Restoration Priorities*. Issues in Ecology Report 16. Ecological Society of America, Washington, DC, USA.

Russi, D., Pantzar, M., Kettunen, M., Gitti, G., Mutafoglu, K. et al. (2016) *Socio-Economic Benefits of the EU Marine Protected Areas*. Report prepared by the Institute for European Environmental Policy (IEEP) for DG Environment. IEEP, London, UK.

Sadio, O., Simier, M., Ecoutin, J.-M., Raffray, J., Laé, R. and Tito de Morais, L. (2015) Effect of a marine protected area on tropical estuarine fish assemblages: comparison between protected and unprotected sites in Senegal. *Ocean and Coastal Management* 116: 257–269.

Salcone, J., Brader, L. and Seidl, A. (2016) *Guidance Manual on Economic Valuation of Marine and Coastal Ecosystem Services in the Pacific*. Report of the MACBIO project. (GTZ, IUCN, SREP). Suva, Fiji.

Salm, R.V., Clark, J.R. and Siirila, E. (2000) *Marine and Coastal Protected Areas: A Guide for Planners and Managers*. IUCN, Washington, DC, USA.

Sanchirico, J.N., Smith, M.D. and Lipton, D.W. (2008) An empirical approach to ecosystem-based fishery management. *Ecological Economics* 64(3): 586–596.

SCBD and NCEA (2006) *Biodiversity in Impact Assessment*. Background Document to CBD Decision VIII/28 on Voluntary Guidelines on Biodiversity-Inclusive Impact Assessment. Published by the Secretariat of the Convention on Biological Diversity and the Netherlands Commission for Environmental Assessment. Montreal, Canada. CBD Technical Series 26. Available at: www.cbd.int/doc/publications/cbd-ts-26-en.pdf (accessed 30 July 2025).

Sciberras, M., Jenkins, S.R., Mant, R., Kaiser, M.J., Hawkins, S.J. and Pullin, A.S. (2015) Evaluating the relative conservation value of fully and partially protected marine areas. *Fish and Fisheries* 16(1): 58–77.

Sekercioglu, C.H. (2010) Ecosystem functions and services. In: Sodhi, N.S. and Ehrlich, P.R. (eds) *Conservation Biology for All*. Oxford University Press, Oxford, UK.

Shackell, N., Keith, D.M., and Lotze, H.K. (2021) Challenges of gauging the impact of area-based fishery closures and OECMs: a case study using long-standing Canadian groundfish closures. *Frontiers in Marine Science* 8.

Shepherd, G. (2004) *The Ecosystem Approach: Five Steps to Implementation*. IUCN, Gland, Switzerland.

Shepherd, G. (ed.) (2008) *The Ecosystem Approach: Learning from Experience*. IUCN, Gland, Switzerland.

Shin, Y.-J., Bundy, A., Shannon, L.J., Simier, M., Coll, M. *et al.* (2010) Can simple be useful and reliable? Using ecological indicators to represent and compare the states of marine ecosystems. *ICES Journal of Marine Science* 67: 713–731.

Smessaert, J., Missemer, A. and Levrel, H. (2020) The commodification of nature, a review in social sciences. *Ecological Economics* 172: 106624.

Smith, A.D.M., Fulton, E.J., Hobday, D.C., Smith, D.C. and Shoulder, P. (2007) Scientific tools to support the practical implementation of ecosystem-based fisheries management. *ICES Journal of Marine Sciences* 64(4, Special Issue): 633–639,

Smith, J.G., Free, C.M., Lopazanski, C., Brun, J., Anderson, C.R. *et al.* (2023) A marine protected area network does not confer community structure resilience to a marine heatwave across coastal ecosystems. *Global Change Biology* 29: 5634–5651.

Spalding, M., Meliane, I., Bennet, N.J., Dearden, P., Pati, P. and Brumbaugh, R.D. (2016) Building towards the marine conservation end-game: consolidating the role of MPAs in a future ocean. *Aquatic Conservation: Marine and Freshwater Ecosystems* 26(Suppl. 2): 185–199.

Squires, D. and Garcia, S.M. (2018) The least-cost biodiversity impact mitigation hierarchy with a focus on marine fisheries and bycatch issues. *Conservation Biology* 35(2): 989–997.

Standing, A. (2024) Why the $700 Billion Funding Gap for Biodiversity is Dangerous Nonsense: Implications for the Oceans and Small-scale Fisheries. Coalition for Fair Fisheries Arrangements Policy Brief. Available at: www.cffacape.org/publications-blog/funding-gap-dangerous-nonsense (accessed 30 July 2025).

STECF (2007) Working Group Report on Evaluation of closed Areas Schemes. ISPRA, Italy, 5–9 November. Available at: https://marine.gov.scot/sma/content/working-group-report-evaluation-closed-area-schemes-sgmos-07-03-ispra-15-19-october-2007 (accessed 30 July 2025).

Stein, D. and Walters, C. (2012) *Understanding Theory of Change in International Development*. Justice and Security Research Programme (JSRP), Asia Foundation. Available at: https://eprints.lse.ac.uk/56359 (accessed 30 July 2025).

Tengö, M., Hill, R., Malmer, P., Raymond, C.M., Spierenburg, M. *et al.* (2017) Weaving knowledge systems in IPBES, CBD and beyond – lessons learned for sustainability. *Current Opinion in Environmental Sustainability* 26–27: 17–25.

ten Kate, K. and Crowe, M.L.A. (2014) *Biodiversity Offsets: Policy Options for Governments*. IUCN, Gland, Switzerland. Available at: https://portals.iucn.org/library/sites/library/files/documents/2014-028.pdf (accessed 30 July 2025).

Thermes, S., Van Anrooy, R., Gudmundsson, A. and Davy, D. (2023) Classification and Definition of Fishing Vessel Types, 2nd edn. FAO Fisheries and Aquaculture Technical Paper 267. FAO, Rome, Italy.

Thomas, C.D. and Gillingham, P.K. (2015)The performance of protected areas for biodiversity under climate change. Biological Journal of the Linnean Society 115: 718–730.

Todd, C., Stevenson, B.N. and Tissot, W.J. (2013) Socioeconomic consequences of fishing displacement from marine protected areas in Hawaii. *Biological Conservation* 160: 50–58.

Tracey, S., Buxton, C., Gardner, C., Green, B., Hartmann, K. *et al.* (2013) Super trawler scuppered in Australian fisheries management reform. *Fisheries* 38(8): 345–350.

Trenkel, V.M., Rochet, M-J. and Mesnil, B. (2007) From model-based prescriptive advice to indicator-based interactive advice. *ICES Journal of Marine Science* 64: 768–774.

UNEP (2014) *Guidance Manual on Valuation and Accounting of Ecosystem Services for Small Island Developing States.* Regional Seas Reports and Studies 193. Available at: www.cbd.int/financial/monterreytradetech/unep-valuation-sids.pdf

UNEP-WCMC (2019) *User Manual for the World Database on Protected Areas and world database on other effective area-based conservation measures: 1.6.* UNEP-WCMC, Cambridge, UK. Available at: http://wcmc.io/WDPA_Manual (accessed 30 July 2025).

United Nations (1992a) *Earth Summit Agenda 21.* United Nations, New York, USA.

United Nations (1992b) *Convention on Biological Diversity.* United Nations, New York, USA.

United Nations (2005) *Report of the International Workshop on Methodologies regarding Free, Prior and Informed Consent and Indigenous Peoples.* 17–19 January, New York.

United Nations (2007a) *United Nations Declaration on the Rights of Indigenous Peoples.* Available at: www.un.org/development/desa/indigenouspeoples/wp-content/uploads/sites/19/2018/11/UNDRIP_E_web.pdf (accessed 30 July 2025).

United Nations (2007b) Resolution 61/105 on sustainable fisheries, including through the 1995 Agreement for the Implementation of the Provisions of the United Nations Convention on the Law of the Sea of 10 December 1982 relating to the Conservation and Management of Straddling Fish Stocks and Highly Migratory Fish Stocks, and related instruments. Available at: https://docs.un.org/en/A/RES/61/105 (accessed 30 July 2025).

Vierros, M., Cresswell, I., Escobar Briones, E., Rice, J. and Ardron, J. (2009) *Global Open Oceans and Deep Seabed (GOODS): Biogeographic Classification.* IOC Technical Series 84 (201). IOC, Paris, France.

Visconti, P., Butchart, S.H.M., Brooks, T.M., Langhammer, P.F., Marnevick, D. *et al.* (2019) Protected area targets post 2020. *Science Policy Forum* 364: 239–241.

Vivacqua, M. (2018) Coastal-marine extractive reserves: reflections on the pre-implementation stage. *Ambiente & Sociedade* 21. Available at: https://doi.org/10.1590/1809-4422asoc0032r3vu18L1AO (accessed 30 July 2025).

Walters, C.J. (1980) Systems principles in fisheries management. In: Lackey, R.T. and Nielsen, L.A. (eds) *Fisheries Management.* John Wiley, New York, USA, pp. 167–183.

Walters, C.J. and Hilborn, R. (1976) Adaptive control of fishing systems. *Journal of the Fisheries Research Board of Canada* 33(1): 145–159.

Wang, S., Kenchington, E., Murillo, F.J., Lirette, C., Wang, Z. *et al.* (2024) Quantifying the effects of fragmentation of connectivity networks of deep-sea vulnerable marine ecosystems. *Diversity and Distributions* 30: e13824.

Ward, T.J., Heinemann, D. and Evans, N. (2001) *The Role of Marine Reserves as Fisheries Management Tools: a Review of Concepts, Evidence and International Experience.* Bureau of Rural Sciences, Canberra, Australia.

Watkins, G., Atkinson, R., Canfield, E., Corales, D., Dixon, J. *et al.* (2015) *Guidance for Assessing and Managing Biodiversity Impacts and Risks in Inter-American Development Bank Supported Operations.* IDB Technical Note 932. Available at: https://publications.iadb.org/en/guidance-assessing-and-managing-biodiversity-impacts-and-risks-inter-american-development-bank (accessed 30 July 2025).

Wauchope, H.S., zu Ermgassen, S.O.S.E., Jones, J.P.G., Carter, H., Schulte to Bühne, H. and Milner-Gulland, E.J. (2024) What is a unit of nature? Measurement challenges in the emerging biodiversity credit market. *Proceedings of the Royal Society B* 291: 20242353.

Weigel, J.Y., Mannle, K.O., Bennet, N.J., Carter, E., Westlund, L. *et al.* (2014) Marine protected areas and fisheries: bridging the divide. *Aquatic Conservation: Marine and Freshwater Ecosystems* 24(Suppl. 2): 192–215.

Westlund, L., Charles, A., Garcia, S.M. and Sanders, J. (2017) *Marine Protected Areas: Interactions with Fishery Livelihoods and Food Security.* FAO Fisheries and Aquaculture Technical Paper 603. FAO, Rome, Italy.

White, J.W., Kilduff, D.P., Hastings, A. and Botsford, L.W. (2024) Marine reserves can buffer against environmental fluctuations for overexploited but not sustainably harvested fisheries. *Ecological Applications* 34: e3043.

Wilcox, C. and Donlan, C.J. (2007) Compensatory mitigation as a solution to fisheries bycatch-biodiversity conservation conflicts. *Frontiers in Ecology and the Environment* 5: 325–331.

Williams, K.J., Harwood, T.D. and Ferrier, S. (2016) *Assessing the Ecological Representativeness of Australia's Terrestrial National Reserve System: a Community-level Modelling Approach*. Available at: https://publications.csiro.au/rpr/pub?pid=csiro:EP163634 (accessed 30 July 2025).

Willer, D.F., Brian, J.I., Derrick, C.J., Hicks, M., Pacay, A. *et al.* (2022) 'Destructive fishing' – a ubiquitously used but vague term? Usage and impacts across academic research, media and policy. *Fish and Fisheries* 23: 1039–1054.

Wilson, D.C. and Delaney, A.E. (2005) Scientific knowledge and participation in the governance of fisheries in the North Sea. In: Gray, T.S. (ed.) *Participation in Fisheries Governance*. Springer, Dordrecht, The Netherlands, pp. 319–341.

Whitty, T.S. (2023) Oceans and communities. In: Obaidullah, F. (ed.) *The Ocean and Us* Springer, Cham, Switzerland, pp. 257–265.

Worboys, G.L. (2010) The connectivity conservation imperative. In: Worboys, G.L., Francis, W.L. and Lockwood, M. (eds) *Connectivity Conservation Management. A Global Guide*. Earthscan, London,UK, pp. 3–21.

WRI (2000) *Fish for the Future? Testing Conservation Assumptions behind a Portfolio of Small-scale Marine Reserves in the Indo-Pacific – the Sulu-Celebes Region*. World Resources Institute. Workshop held at the Sarabia Manor, Iloilo City, The Philippines, 15–18 August.

Wright, G., Gjerde, K.M., Johnson, D.E., Finkelstein, A., Ferreira, A.M. *et al.* (2018) *Marine Spatial Planning in Areas Beyond National Jurisdiction*. Issue Brief 08/18. IDDRI, Paris, France.

Young, O.R., Osherenko, G., Ekstrom, J., Crowder, L.B., Ogden, J. *et al.* (2007) Solving the crisis in ocean governance. Place-based management of marine ecosystems. *Environment* 49(4): 20–32.

Zhou, S. and Griffiths, S.P. (2008) Sustainability Assessment for Fishing Effects (SAFE): a new quantitative ecological risk assessment method and its application to elasmobranch bycatch in an Australian trawl fishery. *Fisheries Research* 91: 56–68.

Zhou, S. and Smith, A.D.M. (2017) Effect of fishing intensity and selectivity on trophic structure and fishery production. *Marine Ecology Progress Series* 585: 185–198.

Zhou, S., Smith, A.D.M. and Fuller, M. (2011) Quantitative ecological risk assessment for fishing effects on diverse data-poor non-target species in a multi-sector and multi-gear fishery. *Fisheries Research* 112(3): 168–178.

Zhou, S., Hobday, A.J., Dichmont, C.M. and Smith, A.D. (2016) Ecological risk assessments for the effects of fishing: a comparison and validation of PSA and SAFE. *Fisheries Research* 183: 518–529.

Zhou, S., Kolding, J., Garcia, S.M., Plank, M.J., Bundy, A. *et al.* (2019) Balanced harvest: concept, policies, evidence, and management implications. *Reviews in Fish Biology and Fisheries* 29: 711–733.

Zuo, J., Pullen, S., Palmer, J., Bennetts, H., Chileshe, N. and Ma, T. (2015) Impacts of heat waves and corresponding measures: a review. *Journal of Cleaner Production* 92: 1–12.

Appendix 1

Example of Scoring OECM Criteria Using An Expert-based Approach[1]

An expert-based MCDA was undertaken in the Aegean Sea (Petza *et al.*, 2019) to assess 516 fishery restricted areas (FRAS) as potential OECMs. FRAs are closed areas defined by the General Fisheries Council of the Mediterranean (GFCM) as:

> a geographically defined area in which all or certain fishing activities are temporarily or permanently banned or restricted in order to improve the exploitation and conservation of harvested living aquatic resources or the protection of marine ecosystems.

For their study, Petza *et al.* broadened this definition to cover areas closed to fishing by *environmental, archaeological, or maritime legislation [at] national, European... or international... levels.*

Based on the literature available at the time of the analysis, a small group of fisheries and conservation experts identified seven criteria against which potential OECMs could be assessed (Table A1.1, col. 1). These criteria do not match those identified in Decision 14/8 because they were identified by the experts, before the CBD COP Decision was adopted. In addition, because of the broadened definition used, many FRAs overlapped significantly with already designated protected areas (Petza *et al.*, 2019: 6), inadvertently violating the most important criteria of the OECM identification process.[2] It should also be noted that the criteria elicited by the experts were all related to the actions taken in the FRAs (objectives, regulations, governance) and not to their observed or intended biodiversity outcomes. The results are of interest, however, from both historical and methodological points of view.

A MCDA framework was proposed to assess, based on expert views, the extent to which individual potential OECMs would sufficiently contribute to marine biodiversity conservation and hence could be formally identified as OECMs.

In order to set the MCDA framework, a number of rating classes (or properties) were determined by experts (Table A1.1, col. 2) for each criterion, from the information available in the literature and in the FRAs database (Petza *et al.*, 2017). Each rating property was allocated a score from 0 to 100 by each expert, based on its importance for the biodiversity objective, and the median score of the expert group was taken as the consolidated score for the rating property (Table A1.1, col. 3).

Table A1.1. Theoretical example of criteria, rating classes and composite scores elaborated for an expert-based multiple criteria decision analysis of OECMs (based on Petza *et al.*, 2019: supplementary Table 3). The score reached by the criteria (column 3, in bold) and the resulting weighted scores and resulting composite score (column 5) have been added and are only illustrative. Table used with permission from Elsevier.

Criteria	Rating classes (properties)	Scoring	Weight (Tot = 1)	Weighted score
1. The area is geographically well defined:	**By co-ordinates**	**100**	0.03	3.0
	By description	70		
	Not defined	0		
2. The biodiversity conservation objective is to...	Protect biodiversity as a whole	100	0.14	
	Protect specific habitats	**80**		11.2
	Protect specific stocks	60		
	None, but contributes significantly	30		
	None: but contributes slightly	10		
3. Activities allowed within the area to meet biodiversity conservation objectives are...	No fishing activity	100	0.23	
	Static gears only	**60**		13.8
	Mobile gears	50		
	Static and mobile gears	40		
	Towed gears	20		
	Towed and static gears	15		
	Towed and mobile gears	10		
	All gears	5		
4. Management and control mechanisms exist within the area?	Yes, all needed	100	0.37	
	Partially	**50**		18.5
	No	0		
5. Area is in place for the long term	>60 years	100	0.06	
	59–40 years	90		
	39–20 years	**60**		3.6
	19–10 years	40		
	<10 years	20		
6. Mechanisms by which area is established are difficult to reverse	EU legislation	100	0.06	
	RFMOs' decisions	90		
	National law	**80**		4.8
	Presidential/Royal decree	60		
	Joint ministerial decision	40		
	Ministerial decision	20		
7. Area closure during the year is:	**Permanent**	**100**	0.11	11.0
	Seasonal: >240 days/year	60		
	Seasonal: 180–239 days/year	40		
	Seasonal: 1–179 days/year	20		
Composite score	**Moderately effective**			**65.9/100**

Independently, the seven criteria were also weighted and ranked by the experts, using the Analytic Hierarchy Process (AHP) based on pairwise comparison of the criteria.

For the case-by-case implementation of the MCDA to each potential OECM, each criterion was initially scored as indicated above, and then the consolidated scores of the seven criteria were aggregated using a weighted additive model to produce the overall composite score of the OECM ranging from 0% to 100% (this is not shown in Table A1.1). Finally, the potential OECMs were classified into six classes of effectiveness according to their composite scores, as follows: (A) extremely effective (composite score from 100% to 90%); (B) very effective (89–80%); (C) effective (79–70%); (D) moderately effective (69–60%); (E) slightly effective (59–50%); and (F) ineffective (<49%). The percentage limits of the classes of effectiveness were expert based. The minimum standard (class of effectiveness) that a *bona fide* OECM must meet might be suggested by the expert group but should be formally decided by the decision makers. The workflow may be followed in Table A1.1.

To check the validity of their expert-based process, Petza *et al.* (2019) undertook an analysis of the consistency of the experts' judgements and a sensitivity analysis.

This type of MCDA might be used as model for addressing the OECMs issue when other types of areas are to be assessed, e.g., using the CBD Decision set of identification criteria (or Steps) and adjusting accordingly the set of rating properties, scoring range and classed of effectiveness.

Notes

[1] Disclaimer: This annex, developed by the authors of this book, based on the original paper by Petza *et al.* (2019), is purely illustrative of an example of a useful multiple criteria scoring process applied to OECMs. This does not imply that the authors endorse the criteria, rating classes, scores and conclusions of the cited analysis. Any error or misinterpretation is our responsibility.

[2] i.e. that areas being considered as potential OECMs should not have been already designated as MPAs.

Appendix 2

Criteria for Identification

Criteria for identification (shaded rows), subcriteria (column 1) and elements of evidence (column 2), based on CBD (2018). These elements are sequentially numbered (A, B, B1a, B1b, etc.) for easier cross-reference in the text.

Criterion A: Area is not currently recognized as a protected area		
A Not a protected area	Aa	The area is not currently recognized or reported as a protected area or part of a protected area.
	Ab	It may have been established for another function.
Criterion B: Area is governed and managed		
B1 Geographically defined space	B1a	Size and area are described, including in three dimensions where necessary.
	B1b	Boundaries are geographically delineated.
B2 Legitimate governance authorities	B2a	Governance has legitimate authority – and is appropriate for achieving *in situ* conservation of biodiversity within the area.
	B2b	Governance by Indigenous Peoples and local communities is self-identified in accordance with national legislation and applicable international obligations.
	B2c	Governance reflects the equity considerations adopted in the Convention.
	B2d	Governance may be by a single authority and/or organization or through collaboration among relevant authorities and provides the ability to address threats collectively.
B3 Managed	B3a	Managed in ways that achieve positive and sustained outcomes for the conservation of biological diversity.
	B3b	Relevant authorities and stakeholders are identified and involved in management.
	B3c	A management system is in place that contributes to sustaining the *in situ* conservation of biodiversity.
	B3d	Management is consistent with the ecosystem approach with the ability to adapt to achieve expected biodiversity conservation outcomes, including long-term outcomes, and including the ability to manage a new threat.

Criterion C: Achieves sustained and effective contribution to *in situ* conservation of biodiversity		
C1 Effective	C1a	The area achieves, or is expected to achieve, positive and sustained outcomes for the *in situ* conservation of biodiversity.
	C1b	Threats, existing or reasonably anticipated ones, are addressed effectively by preventing, significantly reducing or eliminating them, and by restoring degraded ecosystems.
	C1c	Mechanisms, such as policy frameworks and regulations, are in place to recognize and respond to new threats.
	C1d	To the extent relevant and possible, management inside and outside the other effective area-based conservation measure is integrated.
C2 Sustained over long term	C2a	The other effective area-based conservation measures are in place for the long term or are likely to be. 'Sustained' pertains to the continuity of governance and management and 'long term' pertains to the biodiversity outcome.
C3 *In situ* conservation of biological diversity	C3	Recognition of other effective area-based conservation measures is expected to include identification of the range of biodiversity attributes for which the site is considered important (e.g. communities of rare, threatened or endangered species, representative natural ecosystems, range-restricted species, key biodiversity areas, areas providing critical ecosystem functions and services, areas for ecological connectivity).
C4 Information and monitoring	C4a	Identification of other effective area-based conservation measures should, to the extent possible, document the known biodiversity attributes, as well as, where relevant, cultural and/or spiritual values of the area and the governance and management in place as a baseline for assessing effectiveness.
	C4b	A monitoring system informs management on the effectiveness of measures with respect to biodiversity, including the health of ecosystems.
	C4c	Processes should be in place to evaluate the effectiveness of governance and management, including with respect to equity.
	C4d	General data of the area such as boundaries, aim and governance are available information.

Criterion D: Associated ecosystem functions and services and cultural, spiritual, socioeconomic and other locally relevant values		
D1 Ecosystem functions and services	D1a	Ecosystem functions and services are supported, including those of importance to Indigenous Peoples and local communities, for other effective area-based conservation measures concerning their territories, taking into account interactions and trade-offs among ecosystem functions and services, with a view to ensuring positive biodiversity outcomes and equity.
	D1b	Management to enhance one particular ecosystem function or service does not impact negatively on the site's overall biological diversity.
D2 Cultural, spiritual, socioeconomic and other locally relevant values	D2a	Governance and management measures identify, respect and uphold the cultural, spiritual, socioeconomic and other locally relevant values of the area, where such values exist.
	D2b	Governance and management measures respect and uphold the knowledge, practices and institutions that are fundamental for the *in situ* conservation of biodiversity.

Appendix 3

Glossary[1]

Additional properties Important OECM properties referred to in Decision 14/8, such as ecological representativeness, connectivity, complementarity and integration. In this book, they are referred to as 'additional properties' because properties such as ecological representativeness, complementarity with MPAs, connectivity across ecological networks, or integration across seascape are mentioned in Decision 14/8. They are not directly mentioned in the identification criteria but their importance is emphasized in the voluntary guidance. If strongly present, they support the OECM identification case, but if weak or absent do not disqualify an area from being an OECM when the criteria for identification have been adequately met.

Area-based fishery management measure (ABFM) A formally established, spatially defined, fishery management and/or conservation measure, implemented to achieve one or more intended fishery outcomes, commonly related to sustainable use of the fishery. However, it can also often include protection of, or reduction of impact on, biodiversity, habitats or ecosystem structure and function (CBD, 2018).

Area-based management tool (ABMT) A tool, including marine protected areas, used to manage activities or sectors within a geographically defined area to achieve specific conservation and sustainable use objectives. These tools are designed to enhance the conservation and sustainable use of marine biodiversity within a designated area (BBNJ Agreement, Part 1, Article 1).

Benefit An intended positive outcome of a policy or measure, benefitting **biodiversity attributes of concern** (e.g. threatened species, habitats, ecosystem function) and/or human communities (e.g. revenues, livelihood, recreation, spiritual values). The benefit may be a positive change in the feature concerned (e.g. an increase in biomass or a decrease in risk for that biomass), or prevention of a loss that would have occurred if the measure was not in place. All types of benefits may be measured at different spatial scales, e.g. within the OECM, the fishery, the conservation network, or the ecosystem. See also **co-benefit** and **net benefit**.

Biodiversity attributes of concern Biodiversity attributes (or components) in addition to the target species that are: (i) impacted by fishing operations, and for which additional conservation measures are expected to eliminate, reduce or mitigate the impact (particularly the significant adverse impacts, SAIs), and eventually restore healthier conditions; or (ii) identified as a conservation priority, by a legitimate local, national or international authority.

Biogenic habitat Habitat of numerous other species developed by, e.g., coralline algal assemblages (mearl), *Posidonia* beds, kelp forests, oyster beds, hot vents, etc.

By-catch Part of a catch of a fishing unit taken incidentally in addition to the target species towards which fishing effort is directed. Some or all of it may be returned to the sea as discards, usually dead or dying. The term does not include fish released alive under a recreational catch-and-release fishery management.

Candidate-OECM Decision 14/8 is silent on this terminology of OECMs as they go through the identification steps (see Chapter 5) and States may decide on their own terminology, hopefully used consistently across sectors. In this book, a candidate OECM is an **area-based fishery management measure** (ABFM) that has been fully assessed against the OECM criteria, adequately meets these criteria and is submitted to the legitimate authority for final decision.

Co-benefit Unintended benefit obtained accidentally when pursuing an intended one.[2] Unexpected positive outcome of a policy or management measure. In a fishery-OECM, the positive outcome for the broader **biodiversity attributes of concern** obtained when pursuing conventional fishery sustainability objectives regarding the target species and essential habitats. CBD Decision 14/8 suggests that when the ABFM is granted OECM status, it would be useful to reflect recognized but previously unintended benefits in the OECM objectives, *inter alia* to guide performance assessment.

Conservation area and conserved area A conservation area is an area that is protected by law because of its natural beauty, history or importance as a place for animals or plants to live (Cambridge Dictionary). The term 'conserved area' is used in Indigenous and Community Conserved Areas (ICCAs), and is referred to in CBD Decision 14/8 on OECMs, in Annex IV, p.17. There is no internationally agreed definition for 'conserved area'. A number of definitions have been proposed in the scientific and policy literature (Borrini-Feyerabend *et al.*, 2014; Borrini-Feyerabend and Hill, 2015; Jonas and Jonas, 2019; Marnewick *et al.*, 2020; Jonas *et al.*, 2021, 2024). All definitions agree that the term 'conserved areas' includes all conservation areas beyond areas receiving full legal protection from all uses (protected areas). They differ in the degree of protection offered and ways to provide it. In this book, we have used mainly the term **conservation area**.

Discards Target and non-target fish caught and returned to the sea, dead or alive, whether or not brought fully on board a fishing vessel, because of lack of market, lack of space in hull, physical damage and legal requirement regarding *inter alia* minimum size limits or quotas or protected species.

Ecosystem functions Functions or processes contributing to maintain the ecosystem operations and characteristics. They include decomposition, production and nutrient and energy cycling. The diversity of species contributing to each ecosystem function is a key property of each ecosystem, and maintaining functional diversity is a central consideration in conservation and sustainable use. See also **Ecosystem services**.

Ecosystem services The benefits people obtain from ecosystems. These include provisioning services such as food and water production; regulating services such as flood and disease control; cultural services such as spiritual, recreational and cultural benefits; and supporting services such as nutrient cycling that maintain the conditions for life on Earth (Millennium Ecosystem Assessment, 2005). See also **Ecosystem functions**.

Fishery-OECM In this book, a geographically defined area other than a protected area, which is governed and managed by a mandated fisheries authority in ways that achieve positive and sustained long-term outcomes for the *in situ* conservation of biodiversity, with associated ecosystem functions and services and, where applicable, cultural, spiritual, socioeconomic and other locally relevant values, along with any intended fishery outcomes.

Free, prior and informed consent In line with the UN Declaration on Rights of Indigenous Peoples (UNDRIP) (United Nations, 2007), free, prior and informed consent (FPIC) centres on obtaining consent from Indigenous Peoples and, in many jurisdictions, other local communities for any activities undertaken on their land and any use of information they choose to provide to non-band/community members. (www.ihrb.org/resources/what-is-free-prior-and-informed-consent-fpic)

In situ **conservation of biodiversity** The conservation of ecosystems and natural habitats and the maintenance and recovery of viable populations of species in their natural surroundings, and in the case of domesticated or cultivated species, in the surroundings where they have developed their distinctive characteristics (CBD, 1992, Article 2).

Key biodiversity areas (KBA) Sites contributing significantly to the global persistence of biodiversity (IUCN, 2021).

Legitimate authority Authority with the formal mandate (traditional or given by the State) for decision making in an area or sector. In marine fisheries, within national EEZs, the State, or any authority mandated by the State such as a ministry of fisheries. In ABNJ, when they have been developed consistent with UNCLOS, RFMOs/As are a legitimate authority for fisheries within their area of jurisdiction. Depending on the context, reference may also be made to legitimate governance or management authority.

Locally managed marine area (LMMA) An area of nearshore waters with associated coastal and marine resources that is largely or wholly managed at a local level by the coastal communities, land-owning groups, partner organizations and/or collaborative government representatives who reside or are based in the immediate area. (http://lmmanetwork.org)

Long term Decision 14/8 (criterion C2) states that the OECMs and the positive biodiversity outcomes are expected (or likely to be) to be in place for the long term. The 'length' of the term is not defined but the intent – as illustrated by legal, policy or regulatory means or through a public commitment – is to maintain the OECM status and its benefits for as long as possible and necessary, with any changes in the measures requiring consultation and substantive governance actions.

Mainstreaming The process of embedding biodiversity considerations into policies, strategies and practices of key public and private actors that impact or rely on biodiversity, so that it is conserved and sustainably and equitably used both locally and globally (GEF in Huntley and Redford, 2014). Similarly, biodiversity mainstreaming is generally understood as ensuring that biodiversity, and the services it provides, are appropriately and adequately factored into policies and practices that rely and have an impact on it. More specifically, in fisheries, it is the progressive, interactive process of recognizing the values of biodiverse natural systems in the development and management of fisheries, accepting full accountability for, and effectively responding to, the broader impact of fishing and fishery-related activities on biodiversity and related structure and function of ecosystems (Friedman *et al.*, 2018). (www.cbd.int/mainstreaming)

Management authority The authority in charge of the management of resources and their use or protection in an area. It may be the same as the legitimate authority. A parks agency may be a legitimate authority for protected areas, whereas management of uses of natural resources resides with other agencies. In marine fisheries, the management authority is often a ministry or department of fisheries, but could also be a unit of a ministry with broader mandate. An example is the Department for Environment, Food and Rural Affairs (DEFRA), responsible for environmental protection and food production and standards, as well as agriculture, fisheries and rural communities in the entire United Kingdom.

Move-on rule A regulatory provision that requires a fishing vessel that encounters (typically bringing on board during fishing operations) more than a maximum limit of a particular protected taxon or evidence of a vulnerable habitat to move away from the point of encounter, by a minimum regulated distance.

Net benefits In this book, the net gain in the protected biodiversity features of concern in the fishery-OECM, resulting from the fisheries restrictive measures, minus the loss suffered by the same biodiversity features when moving outside the OECM area, caused by the same fishery or other fisheries operating around the OECM. Co-ordinating management inside and outside the OECM should optimize net benefits.

Other effective area-based conservation measure (OECM) A geographically defined area other than a **protected area** which is governed and managed in ways that achieve positive and sustained long-term outcomes for the *in situ* conservation of biodiversity with associated ecosystem functions and services and, where applicable, cultural, spiritual, socioeconomic and other locally relevant values (CBD 2018).

Potential OECM In fisheries, an area where an existing or planned **area-based fishery management measure** (ABFM) is in place or under consideration, that appears, after a quick screening, to have biodiversity and other features considered sufficient enough to justify a full assessment against the CBD Decision 14/8 criteria.

Protected area A geographically defined area, which is designated or regulated and managed to achieve specific conservation objectives (CBD, 1992, Article 2). 'A clearly defined geographical space, recognised, dedicated, and managed, through legal or other effective means, to achieve the long-term conservation of nature with associated ecosystem services and cultural values' (IUCN in Dudley *et al.*, 2008).

Rights-holder In the context of protected areas, 'rights-holders' are actors with legal or customary rights to natural resources and land, in accordance with national legislation (CBD Decision 14/8, p. 7, footnote 23). See also **Stakeholder**.

Stakeholder Generally speaking, a person, group or organization with a vested interest, or stake, in the decision-making process and activities of a given industry, organization or project. Actors with interest and concerns over natural resources and land (CBD Decision 14/8, p. 7, footnote 23). Stakeholders include '**rights-holders**'. In this book we tend to use the term 'stakeholder' to include all individuals and organizations who are considered by the legitimate authority as having an interest and/or a role in the outcome of their OECM decision-making process and activities, regardless of whether or not the legitimate authority has awarded legal rights to their use of the resource. In contexts where distinctions have been made by legitimate authorities in specific uses, we would then follow the use of the authority being cited.

Sustainable use The use of components of biological diversity in a way and at a rate that does not lead to the long-term decline of biological diversity, thereby maintaining its potential to meet the needs and aspirations of present and future generations (CBD Article 2).

Upgradable ABFM Term used in this book to identify an **area-based fishery management measure** (ABFM, fishery closure) that has been assessed as being close to adequately meet the CBD Definition and identification criteria of OECMs but would need some modifications (e.g. in its boundaries or measures applied in it) to meet them adequately. The area remains a potential OECM.

Notes

[1] Note: terms **in bold** in definitions are terms also defined in this glossary.
[2] For example, the protection of coral and sponge reefs in deep-sea vulnerable marine ecosystems (VMEs) in the North Atlantic also produces co-benefits for many species hosted in these habitats.

Appendix 4

OECMs and Fisheries' Scale

A4.1   Introduction

Fishery-OECMs are ABFMs that are granted OECM status because they meet the relevant CBD criteria, particularly those regarding long-term biodiversity benefits. Fishing activities allowed within – and possibly around – their boundaries have always been controlled to ensure the expected benefits for the fishery resources and, in principle, these controls may be expanded to protect biodiversity benefits, regulating the access to the area, the species and habitats to protect, and the type of gears that may be used. Decision 14/8 suggests to:

> identify and prioritize the sectors most responsible for habitat fragmentation, including ... fisheries ... to engage them in developing strategies for mitigating the impacts on protected areas and protected area networks including OECMs... (Annex 1, p. 4e)

and does not refer to the scale of activities of fisheries or any other economic sector that might be allowed in an OECM. Nonetheless, it has been suggested (Day *et al.*, 2012, 2019; IUCN-WCPA, 2019: 6) that 'industrial' fisheries, and indeed any industrial activities, ought to be excluded from MPAs and OECMs as not compatible with their conservation objectives.

Although the compatibility of a given type of fishery activity in a specific fishery-OECM area should be assessed case by case, some generalization would be useful. However, there is no generally agreed definition of 'industrial' fisheries, or size range for their vessels, and there is no simple set of criteria to unambiguously distinguish industrial (or large-scale) fisheries from artisanal (or small-scale) ones. While a 6-metre canoe is clearly different from a 140-metre factory ship, the difficulty seems to be in separating the size range in two clearly distinct categories[1] and different States have made different decisions in this respect (see below).

In relation to OECMs, the issue should mainly be about the level of impact of allowed activities on the OECM biodiversity that might be considered compatible or not with its expected outcomes. It is already internationally agreed that any fishery, at any large or small scale, must be regulated with enhanced risk aversion for significant adverse impacts (SAIs), in general and particularly in vulnerable areas. Moreover, it would seem logical to establish effective fishery-OECMs in all intensively fished areas, in which ABFMs are already operating, irrespective of fishery scale, to mitigate their collateral impact. Currently, restrictions of fishing activities in OECMs remain a decision of the State or legitimate authority, based on objective local information on impacts and objective application of the CBD identification and performance assessment criteria. The following brief considerations may be kept in mind in doing so.

The possible definition of 'industrial' fisheries has been a subject of technical discussions for decades, with no global agreement. The practical distinction between artisanal and industrial fishing varies between countries and socioeconomic contexts (Rousseau *et al.*, 2019).[2] It has also been stated that 'small-scale fisheries are social units with porous boundaries that individual fishers can cross unconsciously or deliberately, blurring the boundaries between the various fisheries' (Wilson and Delaney, 2005).

The terms 'artisanal' and 'industrial' usually refer to the type of business. The term 'artisanal' refers to family or household-based fisheries while the term 'industrial' refers to fisheries operated by corporate companies. Unfortunately, the term 'industrial' has also been used for reduction fisheries producing fish meals and oil. The terms 'large scale' and 'small scale' refer respectively to the gears and vessel size, level of technology and capital investment. Unfortunately, the term 'scale' may also refer sometimes to the geographic expansion of the fishery. The terminology, its use and its implication are examined briefly below.

1. Industrial Fisheries

An industrial fishery is usually a corporate activity, established for profit, with a large capital investment in technology, using large vessels equipped with sophisticated hydraulic, electrical and electronic equipment, and some processing capacity. The vessels can usually stay at sea for a long time, far away from their base port, and consequently crew safety standards require a minimum size ranging from 7 metres in Cape Verde (Knoops, 1995) to 15 metres in the South Pacific islands (Gillett, 2007) or 24 metres in the Cape Town Agreement.[3] The International Maritime Organization (IMO) sets certification requirements of seagoing fishing vessels of 24 metres and above.[4] The FAO (2022) also considers 'large vessels' as those of 24 metres or more, corresponding to a gross registered tonnage (GRT) of 100 tons. Industrial vessels also tend to use more fuel and less manpower per ton of fish caught than small-scale fisheries (SSFs). Capture, preservation and processing may be integrated on board (in factory ships) or through industrial land-based facilities.

From a conservation point of view, industrialized fisheries are usually associated with higher extraction rates, strong examples of persistent and sequential overfishing,[5] capacity to roam among fishing areas if resources are damaged, and stronger environmental impact, particularly on the bottom (e.g. for trawlers). The situation varies greatly among industrial fisheries and regions and generalizations may often be unfair.

2. Artisanal Fisheries

Artisanal fisheries tend to have the opposite properties within a wide range of technological and operational dimensions (Misund *et al.*, 2002; Griffiths *et al.*, 2007). The term 'artisanal' refers to the socioeconomic dimension of the activity. It implies a simple, individual (self-employed) or family type of enterprise, most often operated by the owner or a family member with the support of the household, even though the vessels may sometimes belong to the fishmonger or some external investor. Many are subsistence fisheries (feeding the household and contributing to village food security) but many are significantly commercial, trading catch at national, regional and even global level for high-value species (e.g. from Africa to Europe and Japan). Boat owners may be organized in small co-operatives. Vessels are usually small, undecked with a small crew, and rarely an inboard engine. However, this is not always the case. Vessel length varies: around 15 metres in the UK,[6] 3–18 metres in Nigeria (Inoni and Oyaide, 2007), more than 20 metres in many developed nations (Cochrane and Garcia, 2009) and nowadays up to 25 metres in Senegal (A. Samba, personal communication). However, at international level, the FAO/ILO/IMO guidelines for small decked and undecked fishing vessels consider only vessels below 12 metres (FAO/ILO/ IMO, 2012) and the European Maritime and Fisheries Fund (EMFF) (Regulation 508/2014) defined small-scale fishing as 'carried out by fishing vessels of an overall length of less than 12 metres and not using towed fishing gear'.

From a conservation point of view, SSFs are assumed to produce smaller impacts on biodiversity (because of their smaller size), to be closer to their environment, and more likely to develop community-based stewardship because of their strong dependence on natural resources. This is, however, not always the case.

Between artisanal and industrial fishing activities, there are intermediate categories like 'modern artisanal fisheries' and 'semi-industrial fisheries' with decked vessels of medium size (12–24 metres) and modern equipment such as echo-sounders, inboard engines, powered winches, radio, GPS, etc. but no full-scale onboard processing facilities. This complicates the determination of a cut-off point between artisanal and industrial vessels.

A4.2   Tentative Conclusion

Based on the above elements, it seems that the cut-off point between artisanal (small-scale) and industrial (large-scale) vessel sizes selected by various States and experts may lie between 12 and 25 metres. Any more specific cut-off point may need to be considered in each country or region, considering the empirical distribution of vessel sizes in the total national fleet and eventually other complementary operational criteria, as suggested, for example, by the European Parliament Committee on Fisheries (2012). Finally, it should be stressed that the vessel sizes in a given fleet may change with time, as shown in Senegal (A. Samba, personal communication) for SSFs vessels and by FAO for semi-industrial and industrial vessels (Thermes *et al.*, 2023).

Notes

[1] The literature rfers to artisanal, modern artisanal, semi-industrial and industrial fisheries, reflecting an increase in vessel size and sophistication of the equipment, costs, fishing power, employment and annual harvest.

[2] For example, in the literature, the length of vessels considered as industrial is over 7 m in Cape Verde (Knoops, 1995) but over 15 m in the South Pacific islands (Gillett, 2007). Vessels undoubtedly industrial are around 60 m for squid jigging (www.fao.org/fishery/fishtech/1114/en), over 100 m for catcher/processer factory trawlers, and over 140 m for supergiant trawlers (Tracey et al., 2013). In addition, in the European Union, the term 'industrial fishing' is usually associated with vessels used in fish meal and oil (reduction) fisheries.

[3] https://en.wikipedia.org/wiki/Cape_Town_Agreement?

[4] www.imo.org/en/OurWork/Safety/Pages/Fishing%20Vessels-default.aspx

[5] The successive overfishing of the world stocks, across time and space.

[6] www.gov.uk/government/publications/the-code-of-practice-for-the-safety-of-small-fishing-vessels-of-less-than-15m-length-overall

Index

www.ingramcontent.com/pod-product-compliance
Lightning Source LLC
Chambersburg PA
CBHW052135170526
45162CB00003B/21